Trace Contaminants in Water: Genesis, Rapid Detection and Sustainable Removal Processes

Topical Conference at the 2008 AIChE Annual Meeting

Philadelphia, Pennsylvania
16 - 21 November 2008

ISBN: 978-1-61567-204-2

Trace Contaminants in Water: Genesis, Rapid Detection and Sustainable Removal Processes

Topical Conference at the 2008 AIChE Annual Meeting

Philadelphia, Pennsylvania
16 - 21 November 2008

ISBN: 978-1-61567-204-2

AIChE

Printed from e-media with permission by:

Curran Associates, Inc.
57 Morehouse Lane
Red Hook, NY 12571

Some format issues inherent in the e-media version may also appear in this print version.

Printed by Curran Associates, Inc. (2009)

AIChE
3 Park Avenue
New York, NY 10016-5991

Phone: (203) 702-7660
Fax: (203) 775-5177

www.aiche.org

Additional copies of this publication are available from:

Curran Associates, Inc.
57 Morehouse Lane
Red Hook, NY 12571 USA
Phone: 845-758-0400
Fax: 845-758-2634
Email: curran@proceedings.com
Web: www.proceedings.com

TABLE OF CONTENTS

The Fate, Transport and Control of Persistent Organic ChEmicals In Sediments 1
Danny D. Reible

Synthetic Organic Contaminant Uptake by Carbon Nanotubes 2
James Kilduff, Hyung-Nam Lim

Nanoparticle-Based Functionalized Membrane Reactor for Chloro-Organic Detoxification at Room Temperature: An Overview (Invited Paper) 3
D. Bhattacharyya, Jian Xu, Yit Hong Tee, L. Bachas, David Meyer

In-Situ Reductive Transformation of Chlorinated Solvents and Immobilization of Chromium In Soils Using Polysaccharide-Stabilized Iron Nanoparticles 4
Dongye Zhao, Feng He, Lucida Xu, Christopher B. Roberts

Biosensors for Organophosphate Nerve Agents 5
Ashok Mulchandani

Reduction of Perchlorate and Other Micropollutants in a Hydrogen-Based, Hollow-Fiber Membrane Biofilm Reactor 6
Robert Nerenberg

Arsenic Removal in Bangladesh and South East Asia - the Role of Competitive Oxidation, Sorption and Precipitation Reactions 15
Stephan J. Hug, Ioannis A. Katsoyiannis, Linda C. Roberts, Ralf Kaegi, Andreas Voegelin

Selective Ion Exchange for Removal of Inorganic Trace Contaminants from Water 16
Wolfgang H. Höll

Toxic Heavy Metal Removal by Hollow Fiber Membrane Solvent Extraction-Based Techniques 22
Kamalesh K. Sirkar

Endocrine Disrupting ChEmical Risk Management Research in the US Epa's Office of Research and Development 23
Carolyn Acheson, Marc Mills, Steve Hutchins, Todd Martin, Gregory Sayles, Kathleen Schenk, Douglas Young

Selective and Regenerable Ion Exchange for Perchlorate Removal, Recovery, and Environmental Forensics 24
Bachua Gu

Adsorption of Trace Levels of Arsenic from Aqueous Solutions by Conditioned Layered Double Hydroxides: Flow Experiments and Multi Solute Batch Experiments 25
Megha Dadwhal, Muhammad Sahimi, Theodore T. Tsotsis

Remove Arsenic and MS2 Virus by Iron Oxide Nanoparticles Coated on Glass Fibers 27
Xuan Li, Jinwen Wang, Leonardo A. Gutierrez, Thanh H. Nguyen

Chromium (III) Sequestration by Oats Straw and Agave Baggasse: Sorption Mechanism 28
Refugio Bernardo García Reyes, José René Rangel Méndez

Hybrid Ion Exchanger (HIX): Characterization and New Application Opportunities 33
Arup K. Sengupta

Arsenic Removal from Drinking Water Using a Regenerable Adsorbent 34
Paul Sylvester, Teresia Moller

Investigation of Extraordinarily High Cr(VI) Removal Capacity by a Weak-Base Anion Exchange Resin: The Mechanism of Reactive Ion Exchange 37
Sudipta Sarkar, Nicole Blute, Jin-ChEng Lin, Arup K. SenGupta

Low Cost - Sustainable - High Capacity Potable Water Treatment Media for Arsenic Removal 38
Rajiv Banavali, Jose Antonio Trejo, Garth R. Parker

Rapid Sensing of Toxic Metals In Water Using a Hybrid Inorganic Material 44
Prasun K. Chatterjee, Arup K. Sengupta

Magnetic Edta – Attaching Metal ChElators to Nanomagnets for Water Purification and Heavy Metal Recovery 45
Fabian M. Koehler, Michael Rossier, Evagelos K. Athanassiou, Robert N. Grass, Wendelin J. Stark

Physicochemical Denitrification Process for Drinking Water Resources at Ambient Conditions 46
Masamichi Tsuji, Mitsuo Kawamura, Harue Tsuji

Destruction of Trichloroethylene with Zerovalent Iron/ Persulfate 64
Anita Padmanabhan, John Bergendahl

Relationship Between Micellar and Hemi-Micellar Processes and the Bioavailability of Surfactant-Solubilized Hydrophobic Organic Compounds 65
Derick G. Brown

Metal Oxide Nanoparticles Coated on a Glass Fiber Substrate to Remove Viruses from Water 66
Xuan Li, Leonardo A. Gutierrez, Thanh H. Nguyen, James Economy

Comparison of Treatment Kinetics for Removal of Btex Compounds and Chlorinated Solvents Using Advanced Oxidation Processes 67
Robert W. Peters, M. P. Sharma, Yusuf G. Adewuyi

Disinfection Byproducts Control with Ion Exchange 76
Francis Boodoo, Joe D'Alessandro, Edward T. Begg, Dj Shannahan

Enhanced Chlorinated Organic Removal from Water Using Iron and Bimetallic Nanoparticle Composite Membranes 86
Stephen M. Ritchie

Investigation of Quality Concerns Due to Elastomeric Compunds Degradation in Chloramine-Disinfected Waters 87
Raja Mohan Nagisetty, Thomas D. Rockaway, Gerold A. Willing

Adsorption of Chloroform & MTBE on Granular Zeolites; Batch and Column Adsorption Studies 88
Laila Abu-Lail, John Bergendahl, Robert Thompson, Arjan Giaya

Cost Estimation of Advanced Oxidation Processes (AOPs) Involving Ultrasound. A Comparative Study 89
Naresh N. Mahamuni, Yusuf G. Adewuyi, Robert W. Peters

Advanced Oxidation of Pharmaceuticals: Preparing for Indirect and Direct Water Reuse 90
Lee M. Blaney, Lynn E. Katz, Desmond F. Lawler

Removal of Volatile Organic Compounds from Water by Use of Microcapsules Containing PDMS 91
Ziying Xiang, Yangcheng Lu, Xingchu Gong, Guangsheng Luo

Water Compatible Resin for Separation and Recovery of Dissolved Precious Metals 92
Yu Lung Lam, Kwong Yu Chan, Patrick H Toy

Author Index

50a The Fate, Transport and Control of Persistent Organic Chemicals In Sediments

Danny D. Reible, University of Texas, EWRE C1786, Austin, TX 78712

Contaminated sediments represent some of the most difficulty and potentially costly environmental problems. Contaminated sediment sites are typically characterized by large volumes of marginally contaminated sediments that are made more difficult to manage due to poor access and limited treatment options. Effective management requires a solid understanding of the processes giving rise to exposure and risk as well as recognition of available but limited treatment options that exist. This presentation will summarize the unique characteristics of sediment that control the risk associated with contaminants, current research to develop improved assessment approaches and the opportunities and challenges of management options. The presentation will focus on the role and characteristics of the biologically active zone of sediments and the use and measurement of interstitial water concentrations to assess risk and develop conceptual site models. The limitations of dredging as cited by a recent National Research Council report and conventional passive capping will be summarized. Experiences with approaches providing more effective in-situ management and control of contaminant mobility and risk will be identified.

50b Synthetic Organic Contaminant Uptake by Carbon Nanotubes

James Kilduff and Hyung-Nam Lim. Civil and Environmental Engineering, Rensselaer Polytechnic Institute, 110 8th Street, Troy, NY 12180

Carbon nanotubes (CNTs) hold promise to offer several potential advantages over the existing state of the art adsorbents, activated carbons and carbon fibers. To exploit this promise, purification to remove amorphous carbon and metals is often done. In addition, surface treatment such as oxidation to etch nanotube surfaces, to remove endcaps and expose inner surface area, or to add hydrophilic functional groups to enhance dispersion in aqueous solution, is also done. The effects of such treatments will be assessed in terms of their effects of organic contaminant uptake. We hypothesize that pi-pi interactions may be important in the sorption of aromatic organics, and we test this by comparing a suite of solutes have different numbers of aromatic rings, pi-bonds, and functional groups that withdraw or donate electron density. We present a side-by-side comparison with existing adsorbents, especially activated carbon, to provide a rational basis for selecting and preparing CNTs, and to evaluate whether CNTs deserve further consideration for water treatment. Of particular interest in this regard is whether competition form natural organic matter is mitigated during uptake by CNTs as compared to porous activated carbons.

50c Nanoparticle-Based Functionalized Membrane Reactor for Chloro-Organic Detoxification at Room Temperature: An Overview (Invited Paper)

D. Bhattacharyya[1], Jian Xu[2], Yit Hong Tee[1], L. Bachas[3], and David Meyer[4]. (1) Chemical and Materials Engineering, University of Kentucky, 177 F. Paul Anderson Tower, Lexington, KY 40506, (2) Chemical & Materials Engineering, University of Kentucky, 177 F. Paul Anderson Tower, Lexington, KY 40506, (3) Dept. of Chemistry, University of Kentucky, Lexington, KY 40506, (4) University of Kentucky, 177 Anderson Hall, Lexington, KY 40506

The study of reductive dechlorination technology involved the synthesis of hybrid membrane materials containing bimetallic (Fe/Ni, Fe/Pd) iron-based nanoparticles using either functionalization and ion-exchange principles or phase-inversion membrane processing. We synthesized and immobilized bimetallic nanoparticles with controlled diameters $<$ 30 nm using membrane-based supports (such as PVDF and Chitosan). In addition to the rapid degradation (by Fe/Ni) of TCE (trichloroethylene) to ethane, we were also able to achieve complete dechlorination of selected chloro-biphenyls (PCBs) using milligram quantities immobilized Fe/Pd nanoparticles in membrane domain. For the PVDF membrane supports we synthesized core/shell Fe/Pd nanoparticles in polyvinylidene fluoride (PVDF) microfiltration membranes functionalized with poly(acrylic acid) (PAA). PAA functionalization was achieved by in situ free radical polymerization of acrylic acid in microfiltration membrane pores. Target ferrous ions were then introduced into the membranes by the ion exchange process. Subsequent reduction resulted in the in situ formation of 20-40 nm Fe nanoparticles. Bimetallic nanoparticles can be formed by post-deposition of Pd. The membranes and Fe/Pd nanoparticles were characterized by thermogravimetric analyzer (TGA), FTIR, scanning electron microscopy (SEM), and transmission electron microscopy (TEM). 2,2′-Dichlorobiphenyl (PCB4) and 3,3′,4,4′-tertrachlorobiphenyl (PCB77) were chosen as the model compounds to investigate the catalytic properties of bimetallic nanoparticles, the reaction mechanism, and the intrinsic kinetics. A two-dimensional steady-state model was also developed to correlate and simulate mass transfer and reaction in the membrane pores under pressure-driven convective flow conditions. This research was supported by the NIEHS-SBRP Program.

50d In-Situ Reductive Transformation of Chlorinated Solvents and Immobilization of Chromium In Soils Using Polysaccharide-Stabilized Iron Nanoparticles

Dongye Zhao[1], Feng He[2], Lucida Xu[2], and Christopher B. Roberts[3]. (1) Department of Civil Engineering, Auburn University, Auburn, AL 36832, (2) Civil Engineering, Auburn University, Auburn, AL 36832, (3) Chemical Engineering, Auburn University, Auburn, AL 36832

Remediation of soils and sediments contaminated with chlorinated solvents and heavy metals has been a major environmental challenge for decades. Yet, cost-effective in situ remediation technology remains lacking. While zero-valent iron (ZVI) nanoparticles have been found effective in reducing various chlorinated hydrocarbons, rapid agglomeration of the particles rendered the particles undeliverable in soils. To address this issue, we developed a particle stabilization strategy using starch or carboxymethyl cellulose (CMC) as a stabilizer. We found that the use of the stabilizers can facilitate controlling the size, delivery and transport of the nanoparticles and resulted faster reaction rate. The stabilized ZVI nanoparticles can be readily delivered to the targeted contaminated zones, and can in situ effectively destroy chlorinated solvents such as trichloroethylene (TCE). Bench- and field scale experimental data showed that the stabilized ZVI nanoparticles can in situ completely and rapidly dechlorinate TCE in water and soils. Field tests also indicated that the application of stabilized ZVI nanoparticles can boost long-term biodegradation of chlorinated solvents. Based on the stabilized ZVI nanoparticles, we also developed an technology for in situ reductive immobilization of Cr(VI) in soils and groundwater. When a Cr(VI)-laden soil column was treated with 5.7 bed volumes of 0.06 g/L of the nanoparticles at pH 5.60, only 4.9% of the total Cr was eluted compared to 12% for untreated soil under otherwise identical conditions. Moreover, the ZVI treatment reduced the TCLP leachability of Cr in the soil by 90%, and the California WET leachability by 76%. The stabilized nanoparticles may offer a powerful alternative for in situ dechlorination or in situ reductive immobilization of redox-sensitive heavy metals.

50e Biosensors for Organophosphate Nerve Agents

Ashok Mulchandani, Department of Chemical and Environmental Engineering, University of California, Riverside, CA 92521

Organophosphates, amongst the most toxic substances known, are used widely as pesticides, insecticides and chemical warfare agents. These compounds are potent acetylcholinesterase (AChE) inhibitors. Since AChE is present in all vertebrates, the potential damage caused by OPs to non-target organisms, including humans, is extremely high. Therefore, there is an urgent need for selective, sensitive and rapid analytical devices to monitor these compounds in the field.

This presentation will review our research on biosensors based on organophosphorus hydrolase (OPH), an enzyme that catalyzes the hydrolysis of organophosphorus nerve agents including nerve gases and pesticides, and genetically engineered cells expressing the OPH on its cell surface. These biological recognition elements were integrated with potentiometric, amperometric, optical and conductometric transducers to build field-deployable enzyme- and microbial-biosensors for sensitive, selective, rapid and direct determination of organophosphate pesticides and nerve gases. The enzyme was also used for on-chip precolumn hydrolysis followed by electrophoretic separation and contactless conductivity detection of hydrolysis products to screen and fingerprint the organophosphates.

Reduction of Perchlorate and Other Micropollutants in a Hydrogen-Based, Hollow-Fiber Membrane Biofilm Reactor

Robert Nerenberg, University of Notre Dame, Notre Dame, IN

Introduction

The membrane biofilm reactor (MBfR) a novel bioreactor that provides gaseous substrates directly to a biofilm growing on the membrane surface, avoiding the need for sparging. MBfRs are *not* membrane bioreactors (MBRs), where membranes separate suspended solids from the effluent water, substituting for a clarifier. In MBfRs, a gaseous substrate moves across the membrane. Since the MBfR membranes typically are microporous and hydrophobic, water and bacteria do not penetrate and block the pores. This paper describes H_2-based MBfRs for removal of perchlorate and other oxidized micropollutants contaminants.

Membrane Biofilm Reactors

Configuration

One of the key elements of the MBfR is the membrane. Membranes may be made from organic or inorganic materials, and can be configured in sheet or hollow-fiber geometries. Hollow-fiber membranes are commonly used for MBfRs because, with outside diameters as small as 0.1 mm, they provide high surface-to-volume ratios. Hydrophobic materials are preferred because their pores remain dry, and gas molecules diffuse much more quickly through dry pores than through liquid-filled pores (Yang and L. 1986). Dry pores also eliminate the potential for pore fouling.

Hydrophobic, microporous membranes can be operated at high gas pressures without bubbling. When membrane pores are fairly large, such as with silicon membranes, bubbles form when the gas pressure slightly exceeds the hydrostatic pressure of the surrounding liquid (Ahmed and Semmens 1992; Mulder 1997). In contrast, when the pores are small, the water surface tension provide a significant resistance to bubble formation, allowing much higher applied pressures. Higher gas pressures are advantageous, as they allow greater gas fluxes into the biofilm, providing higher rates of biodegradation.

Figure 1 shows a schematic of a bundle of hollow fiber membranes, and a section of a single hollow fiber. As shown on the right side of the figure, the fibers are collected into a gas-supplying manifold at one end and are sealed at the opposite end. On the left side of the figure, pressurized gas in the lumen (interior) of the fiber diffuses through the dry pores and into the biofilm coating the fiber.

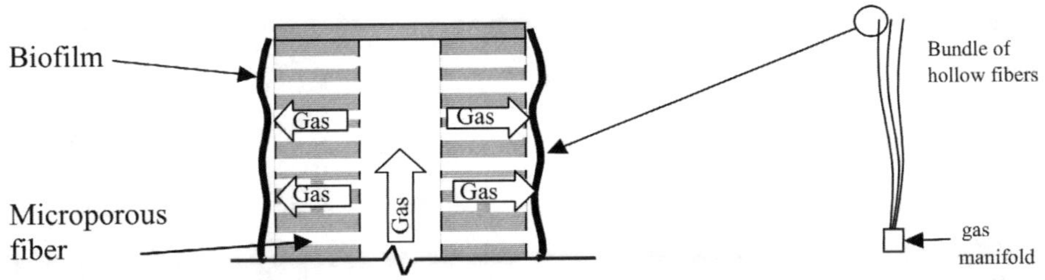

FIGURE 1. Section of fiber (left) and schematic of hollow fiber membrane bundle (right)

Figure 2a shows a scanning electron micrograph (SEM) of the outer surface of a Mitsubishi-Rayon MHF200TL membrane. The average pore diameter is around 0.15 µm, with an elliptical shape, and the surface texture is irregular. Figure 2b is a confocal laser scanning microscopy (CLSM) image showing the cross section of a biofilm growing on an MBfR membrane. In this figure, the biofilm thickness is approximately 50 µm.

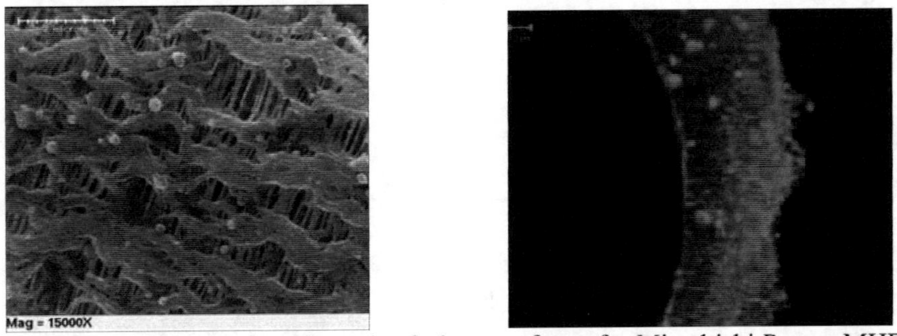

FIGURE 2. (a) Pore structure on the polyethylene surface of a Mitsubishi Rayon MHF200TL16 composite hollow fiber membrane (b) Microscopy image (CLSM) of biofilm growing on hollow-fiber membrane. The red dots are bacteria forming the biofilm, and the outer wall of the membrane (not visible) is immediately left of the biofilm.

Biofilms accumulate naturally on the membrane surface, which is the interface between the gaseous substrate and the dissolved substrate coming from the bulk liquid. Unlike common biofilm applications, where the electron donor and acceptor both diffuse into the biofilm from the bulk liquid (Figure 3a), in an MBfR, one substrate diffused into the biofilm from the membrane and the other from the bulk liquid. This establishes counter-gradients between donor and acceptor (Essila, Semmens et al. 2000; Lee and Rittmann 2002), as shown in Figure 3b. An advantage of counter gradients is that the gaseous substrate supplied from the membrane is "sheltered" from loss into the bulk liquid by the biofilm and by the liquid diffusion layer. Note the high bulk-liquid gas concentration required in Figure 3a, for conventional biofilms, and the much lower bulk-liquid concentration for the MBfR in Figure 3b. If the MBfR gas supply pressure is selected appropriately, very little or no gaseous substrate is lost to the bulk liquid, enhancing the cost-effectiveness of the process.

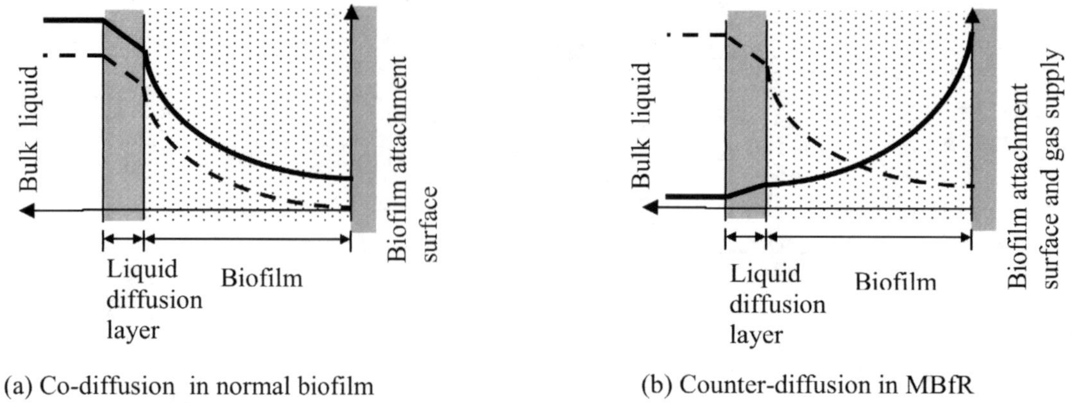

Figure 3. (a) Substrate "co-diffusion" in a normal biofilm and (b) substrate "counter-diffision" in a hollow-fiber membrane biofilm. The bold line is the dissolved gas concentration, while the dotted line is the substrate from the bulk liquid.

Advantages of MBfRs

A major advantage of the MBfR is that virtually all of the gas passing through the membrane can be utilized within the biofilm. This is due to the counter-current transport of dissolved gas and substrate from the bulk liquid, discussed above. Nearly 100% use of the gas means no unproductive loss of gas with the effluent. It also avoids potential hazards with explosive or toxic gases.

A second advantage is that the gas moves across the membrane wall when the bacterially catalyzed reaction in the biofilm creates a concentration gradient. If the biochemical demand for dissolved gas declines, the gradient of and demand for gas also decline. If the demand increases, the gradient and demand also increase. Thus, to a certain degree the MBfR operates as a self-regulating, on-demand system that modulates its gas supply rate to the contaminant load. This prevents wasting gas or having an under-supply.

A third advantage is that hollow-fiber membranes provide a large specific surface area for biofilm accumulation. A high specific surface area allows a high density of contaminant-reducing bacteria in the MBfR. This means that the detention time for the reactor can be small, thereby minimizing capital costs and the system's footprint. It is ideal for treatment-plant retrofits, as well as for new construction.

Hydrogen-Based MBfRs

One of the most exciting applications of the MBfR is to deliver hydrogen gas (H_2) as an electron donor. Many oxidized contaminants can be reduced to less toxic or less mobile species with the addition of an electron donor. The classical example is nitrate, which can be reduced to nitrogen gas by denitrifying bacteria. Historically, organic donors such as methanol, ethanol, and acetate have been used for denitrification. However, H_2 has following inherent advantages over organic electron donors:

- H_2 is a low-cost source of electrons
- H_2 supports autotrophic bacteria, which eliminates the need for an organic carbon source
- H_2 produces less excess biomass (autotrophic growth)
- H_2 cannot leave a significant residuals that increase effluent BOD (low solubility)
- H_2 is non-toxic to humans
- H_2 can be purchased in bulk or generated on-site

Despite these advantages, H_2 has not been widely used, mainly because no efficient and safe delivery system was available. Its low water solubility does not allow it to be supplied in a water stream, and its flammability and cost does not allow H_2 to be sparged.

The MBfR opens the door using H_2 for water and wastewater treatment. In addition to nitrate, a large number of other, relatively new micropollutants fall into the class of being chemically oxidized. Many can be microbially reduced to innocuous or sequestered products. For example:

- Perchlorate (ClO_4^-), a component of solid rocket fuel, can be reduced to Cl^-
- Nitrate (NO_3^-), a common groundwater contaminant, can be reduced to N_2.
- Chlorinated solvents, like trichloroethene (TCE), can be reductively dehalogenated to ethene and Cl^-.
- Bromate (BrO_3^-), an ozonation byproduct, can be reduced to Br^-.
- Selenate (SeO_4^{2-}), which occurs naturally in certain mineral deposits, can be reduced to less mobile selenide (S^{2-}) or elemental selenium (Se°).
- Heavy metals, particularly chromium, can be reduced from hexavalent chromate (CrO_4^{2-}) to less toxic Cr^{3+}.
- Radionuclide metals uranium and neptunium can be reduced to low-mobility U(IV) and Np(IV).

The following section describes experience with H_2-based MBfRs.

H_2-MBfR for Perchlorate in Drinking Water

Perchlorate is an emerging oxidized contaminant in areas affected by military bases and rocket manufacturing and testing. Perchlorate affects thyroid function and is considered an endocrine-disrupting compound. Although no federal standard exist yet, the State of California has an action level of 4 μg/L, and the U.S.E.P.A. anticipates that its health-based standard ultimately will be in the range of 5 - 40 μg/L. Perchlorate can be bacterially respired in a stepwise 8-electron reaction that produces Cl^- ion. The overall reaction is $ClO_4^- + 4H_2 \rightarrow Cl^- + 4H_2O$.

Experiments on a bench-scale MBfR (Figure 4) proved that ClO_4^- could be reliably reduced to below 4 μg/L (Nerenberg and Rittmann 2002; Nerenberg, Rittmann et al. 2002), that the H_2 pressure to the membrane can control the capacity to reduce ClO_4^-, and that prolonged feeding of ClO_4^- enriches the biofilm in perchlorate-reducing bacteria (Nerenberg, Kawagoshi et al. 2008). The bench-scale work also showed that oxygen and nitrate are good electron acceptors to support perchlorate-reducing bacteria, although their concentrations in the MBfR must be very low to preclude inhibition of perchlorate reduction.

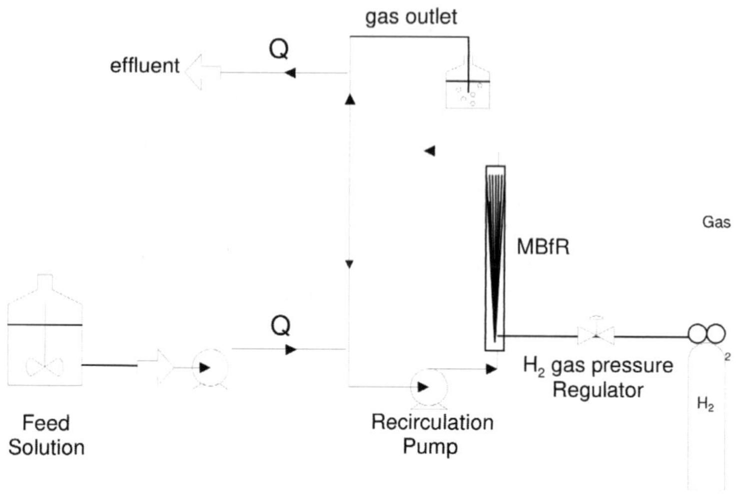

FIGURE 4. Schematic of a bench-scale MBfR

Field-scale pilot testing was carried out at La Puente, California (Nerenberg, Rittmann et al. 2003; Adham, Gillogly et al. 2004). The pilot system consisted of two columns each having ~7,000 hollow-fiber membranes and received a flow rate around 2 L/min. The La Puente groundwater contained approximately 60 μg/L of ClO_4^- and 5.6 mgN/L of NO_3^-. After a start-up period in which practical operating problems were overcome, the pilot-scale system achieved excellent ClO_4^- removal, typically at or below the 4-μg/L action level. Nitrate also was removed to about 0.2 mgN/L, and O_2 was completely removed. One of the most important contributions of the pilot study was quantifying the H_2 use rate, which could not be measured with the small gas flows in the bench-scale studies. The measured H_2 use rate was very close to 100% of the theoretical use rate based on the consumption rate of the three acceptors entering the MBfR: NO_3^-, O_2, and ClO_4^-. The 100% H_2 use means that the MBfR wastes no electron donor, which is essential for good economy, safety, and effluent quality.

The MBfR is currently being commercially developed by Applied Process Technology, Inc. (APT), Pleasant Hill, CA. APT has carried out numerous bench and pilot-scale tests for removal of perchlorate and other oxidized contaminants.

H_2-MBfR for Drinking Water Denitrification

K.-C. Lee (Lee and Rittmann 2000; Lee and Rittmann 2002) was one of the first to study MBfRs for drinking water denitrification. The MBfR was inoculated with a pure culture of the H_2-oxidizing, autotrophic denitrifier, *Ralstonia eutropha,* but was allowed to develop into a mixed culture. Partial denitrification, to 10 mgN/L, was sought and was achieved by limiting the hydrogen supply. Biofilms were allowed to grow to steady state for two different operating conditions. In the first steady state, the influent nitrate was 10 mgN/L, and the membrane hydrogen supply pressure was 0.31 atm (relative to atmospheric pressure). With a hydraulic retention time of 42 minutes, the system achieved 76% nitrate removal and had 0.9 mgN/L nitrite and 0.009 mgH_2/L hydrogen in the effluent. The average biofilm thickness was 110 μm. The second steady-state had an influent nitrate concentration of 12.5 mgN/L and a hydrogen supply pressure of 0.42 atm. In this case, the system achieved 92% nitrate removal and had 0.7 mgN/L nitrite and 0.7 mgH_2/L hydrogen. The biofilm thickness was 179 μm and the effluent biodegradable dissolved organic carbon (BDOC) was 0.5 mgC/L. The final pH was buffered to between 7.1 and 7.2 for both steady-states. The nitrate fluxes were 0.08 and 0.1 $mgNO_3^-N/cm^2$ biofilm surface area/day for the two steady states, respectively. These studies showed that high nitrate fluxes could be achieved with low effluent hydrogen concentrations, thus allowing a compact and efficient process. Also, the degree of nitrate removal was easily controlled by managing the H_2 supply pressure.

H_2-MBfR for Other Oxidized Micropollutants

In addition to perchlorate, several other oxidized micropollutants have emerged as drinking water contaminants, including arsenate ($H_2AsO_4^-$), chromate (CrO_4^{2-}), selenate (SeO_4^{2-}), and bromate (BrO_3^-). In many cases, conventional water treatment processes, as well as oxidative processes such as chlorine-oxidation or ozonation, are ineffective. Advanced separation processes, such as reverse osmosis, ion exchange, membrane filtration, and electrodialysis, can be effective, but are expensive and generate concentrated wastes that require proper disposal. Biological reduction may provide a more suitable treatment alternative, especially when the oxidized contaminant is reduced to a less toxic species (Lovley and Coates 1997).

Many oxidized contaminants are reduced in thermodynamically favorable reactions that have been shown to support bacterial growth. However, in some cases the treatment standards may be below bacterial growth thresholds (S_{min}) (Rittmann and McCarty 2001). In such cases, reduction must occur in parallel to reduction of more amply available "primary" electron acceptors, such as nitrate or oxygen.

Nerenberg and Rittmann (2004) tested a H_2-based MBfR for reduction and removal of several oxidized contaminants when nitrate or oxygen served as primary electron acceptors. The influent concentration of the contaminants was 1 mg/L, while influent oxygen and nitrate were 6 mg/L and 5 mgN/L, respectively. The effluent oxygen and nitrate were below detection. The oxygen reactor had previously been exposed to perchlorate, which explains its higher removals for perchlorate, chlorate, and chlorite. These tests were carried out over a short period of time, without allowing the microbial culture to "adapt" to the new substrate, so long-term efficiencies are likely to be much higher. Results are shown in Table 1.

Table 1. Short-term tests with various oxidized contaminants

Compound	Probable Reduction Reaction(s)	% Removal	
		O_2 Reactor	NO_3^- Reactor
Arsenate	$H_2AsO_4^- + H_2 + H^+ \rightarrow H_3AsO_3 + H_2O$	>50	>50
Bromate	$BrO_3^- + 3H_2 \rightarrow Br^- + 3H_2O$	>95	>95
Chlorate	$ClO_3^- + 3H_2 \rightarrow Cl^- + 3H_2O$	>95	29
Chlorite	$ClO_2^- + 2H_2 \rightarrow Cl^- + 2H_2O$	>75	67
Chromate	$CrO_4^- + 1.5H_2 + 2H^+ \rightarrow Cr(OH)_3$	>75	>75
Dichloro-methane	$DCM + 2\ H_2 \rightarrow CH_4 + 2H^+ + 2\ Cl^-$	38	45
Nitrate	$NO_3^- + 2.5H_2 + H^+ \rightarrow 0.5N_2 + 3H_2O$	Not tested	>99
Perchlorate	$ClO_4^- + 4H_2 \rightarrow Cl^- + 4H_2O$	>98	36
Selenate	$SeO_4^{2-} + 3H_2 + 2H^+ \rightarrow Se^o + 4H_2O$	67	74
Selenite	$HSeO_3^- + 2H_2 + H^+ \rightarrow Se^o + 3H_2O$	93	57

Both reactors showed significant removals for all tested contaminants. Removals ranged from 29% for chlorate in the NO_3^- reactor to over 98% for perchlorate in the O_2 reactor. These results show that many oxidized contaminants can be removed in an MBfR. No specialized inoculum was required, in all cases the required bacteria were present in the mixed culture obtained from an environmental inoculum. Subsequent, more detailed tests have confirmed and expanded the above findings (Chung, Li et al. 2006; Chung, Nerenberg et al. 2006; Chung, Nerenberg et al. 2006; Chung, Ryu et al. 2006; Chung, Nerenberg et al. 2007; Chung and Rittmann 2007; Chung, Rittmann et al. 2007; Downing and Nerenberg 2007).

Oxygen-Based MBfRs

Researchers have also used O_2-based MBfRs for nitrification of drinking water (Brindle and Stephenson 1996) and nitrification and denitrification of wastewater (Syron and Casey 2008). An exciting new strategy is to use the O_2-based MBfR for concurrent nitrification, denitrification, and BOD removal (Semmens, Dahm et al. 2003; Terada, Hibiya et al. 2003; Downing and Nerenberg 2008).

Conclusions

The MBfR is an effective means to deliver gaseous substrates for biological processes, opening the door to a myriad of new applications in water and wastewater treatment. Key applications include nitrification/denitrification and removal of emerging oxidized contaminants. Almost all the supplied gas is delivered to the biofilm, where it is used for desired biochemical reactions, so little waste occurs. The process has a self regulating feature, where increased gas demand from the biofilm creates a greater driving force for gas supply from the membrane. Also, the hollow-fiber membrane configuration also provides high specific surface areas, allowing for compact reactors. The MBfR has been shown to be effective for numerous applications at the bench and pilot scale. Commercial configurations are being developed.

References

Adham, S., T. Gillogly, et al. (2004). "Development of a Membrane Biofilm Reactor (MBfR) for Nitrate and Perchlorate Removal." AWWARF.

Ahmed, T. and M. J. Semmens (1992). "Use of sealed end hollow fibers for bubbleless membrane aeration: experimental studies." Journal of Membrane Science **69**: 1-10.

Brindle, K. and T. Stephenson (1996). "Nitrification in a bubbleless oxygen mass transfer membrane bioreactor." Water Science and Technology **34**(9): 261-267.

Chung, J., X. H. Li, et al. (2006). "Bio-reduction of arsenate using a hydrogen-based membrane biofilm reactor." Chemosphere **65**(1): 24-34.

Chung, J., R. Nerenberg, et al. (2006). "Bio-reduction of soluble chromate using a hydrogen-based membrane biofilm reactor." WATER RESEARCH **40**(8): 1634-1642.

Chung, J., R. Nerenberg, et al. (2006). "Bioreduction of selenate using a hydrogen-based membrane biofilm reactor." Environmental Science & Technology **40**(5): 1664-1671.

Chung, J., R. Nerenberg, et al. (2007). "Evaluation for biological reduction of nitrate and perchlorate in brine water using the hydrogen-based membrane biofilm reactor." Journal of Environmental Engineering-Asce **133**(2): 157-164.

Chung, J. and B. E. Rittmann (2007). "Bio-reductive dechlorination of 1,1,1-trichloroethane and chloroform using a hydrogen-based membrane biofilm reactor." Biotechnology and Bioengineering **97**(1): 52-60.

Chung, J., B. E. Rittmann, et al. (2007). "Simultaneous bio-reduction of nitrate, perchlorate, selenate, chromate, arsenate, and dibromochloropropane using a hydrogen-based membrane biofilm reactor." Biodegradation **18**(2): 199-209.

Chung, J., H. Ryu, et al. (2006). "Community structure and function in a H2-based membrane biofilm reactor capable of bioreduction of selenate and chromate." Applied Microbiology and Biotechnology **72**(6): 1330-1339.

Downing, L. and R. Nerenberg (2007). "Kinetics of Microbial Bromate Reduction in a Hydrogen-Oxidizing, Denitrifying Biofilm Reactor." Biotechnology and Bioengineering **98**(3): 543-550.

Downing, L. and R. Nerenberg (2008). "Total Nitrogen Removal in a Hybrid, Membrane-Aerated Activated Sludge Process. ." Water Research **42**(14): 3697-3708.

Essila, N. J., M. J. Semmens, et al. (2000). "Modeling biofilms on gas-permeable supports: concentration and activity profiles." Journal of Environmental Engineering **126**(3): 250-257.

Lee, K.-C. and B. E. Rittmann (2000). "A novel hollow-fiber membrane biofilm reactor for autohydrogenotrophic denitrification of drinking water." Water Science and Technology **41**(4-5): 219-226.

Lee, K.-C. and B. E. Rittmann (2002). "Applying a novel autohydrogenotrophic hollow-fiber membrane biofilm reactor for denitrification of drinking water." Water Research **36**(8): 2040-2052.

Lovley, D. R. and J. D. Coates (1997). "Bioremediation of metal contamination." Current Opinion in Biotechnology **8**(3): 285-289.

Mulder, M. (1997). Basic principles of membrane technology????

Nerenberg, R., Y. Kawagoshi, et al. (2008). "Microbial Ecology of a Perchlorate-Reducing, Hydrogen-Based Membrane Biofilm Reactor." Water Research **42**(4-5): 1151-1159.

Nerenberg, R. and B. E. Rittmann (2002). "Perchlorate as a Secondary Substrate in a Denitrifying Hollow-Fiber Membrane Biofilm Reactor." Water Science and Technology: Water Supply **2**(2): 259-265.

Nerenberg, R. and B. E. Rittmann (2004). "Reduction of oxidized water contaminants with a hydrogen-based, hollow-fiber membrane biofilm reactor." Water Sci. Technol. **49**(11-12): 223-230.

Nerenberg, R., B. E. Rittmann, et al. (2003). Perchlorate Reduction using a Hollow-Fiber Membrane Biofilm Reactor: Kinetics, Microbial Ecology, and Pilot-Scale Studies. Battelle Bioremdiation Conference 2003, Orlando, FL.

Nerenberg, R., B. E. Rittmann, et al. (2002). "Perchlorate Reduction in a Hydrogen-Based Membrane-Biofilm Reactor." J. AWWA **94**(11): 103-114.

Rittmann, B. E. and P. McCarty (2001). Environmental Biotechnology: Principles and Applications. New York, McGraw-Hill.

Rittmann, B. E., R. Nerenberg, et al. (2004). "Hydrogen-Based Membrane Biofilm Reactor for Wastewater Treatment." War. Sci. Tech. **In press**.

Semmens, M. J., K. Dahm, et al. (2003). "COD and nitrogen removal by biofilms growing on gas permeable membranes." Water Research **37**(18): 4343-4350.

Syron, E. and E. Casey (2008). "Membrane-Aerated Biofilms for High Rate Biotreatment: Performance Appraisal, Engineering Principles, and Development Requirements." Environmental Science & Technology.

Terada, A., K. Hibiya, et al. (2003). "Nitrogen Removal Characteristics and Biofilm Analysis of a Membrane-Aerated Biofilm Reactor Applicable to High-Strength Nitrogenous Wastewater." Journal of Bioscience and Bioengineering **95**(2): 170-178.

Yang, M.-C. and C. E. L. (1986). "Designing hollow-fiber contactors." American Institute of Chemical Engineering Journal **32**(11): 1910-1916.

91b Arsenic Removal in Bangladesh and South East Asia - the Role of Competitive Oxidation, Sorption and Precipitation Reactions

Stephan J. Hug[1], Ioannis A. Katsoyiannis[1], Linda C. Roberts[1], Ralf Kaegi[1], and Andreas Voegelin[2]. (1) Water Resources and Drinking Water, Eawag, Swiss Federal Institute of Aquatic Science and Technology, Ueberlandstr. 133, Duebendorf, CH-8600, Switzerland, (2) Institute of Biogeochemistry and Pollutant Dynamics, ETH Zurich, CHN, Zurich, CH-8092, Switzerland

Over 55% of the 28-35 million Bangladeshi that were initially exposed to arsenic concentrations over 50 ppb still regularly drink water exceeding this limit. The two most successful mitigation options - well switching and installation of deep tube wells - have reached 29% and 12% of the population, respectively. Arsenic removal and four other mitigation options have so far each reached less than 1%, but are likely play a larger role in the future.

The applicability of the various mitigation options is highly dependent on the local conditions.

Arsenic removal with naturally present iron, by aeration and removal of the formed iron(III)(hydr)oxide particles with sorbed arsenic by settling or by sand filtration, is the simplest available water treatment method. Sand filters are now widely used in the red river delta of Vietnam and successfully remove arsenic to below 50 ppb in most cases. In Bangladesh, however, water with low iron and high phosphate concentrations makes arsenic removal difficult. Combined research efforts during the past years have lead to a better understanding of the redox-, sorption-, and precipitation reactions that affect arsenic removal and promising arsenic removal units have been developed. For example, zero-valent iron is successfully used to remove arsenic in tens of thousands of households in Bangladesh (SONO-filters) and Nepal (KANCHAN-filters). However, these filters can fail in certain regions, for reasons which are not yet fully understood. In aerated water, corrosion of iron leads to formation of Fe(II) and the subsequent oxidation of Fe(II) leads to oxidation of As(III), formation of iron(III)(hydr)oxides and coprecipitation and sorption of arsenic and phosphate. Under non-aerated water, iron corrosion leads to anoxic conditions that facilitate sulfate reduction (probably biologically mediated) and to arsenic immobilization by reductive pathways (e.g. incorporation of arsenic into iron corrosion products and to formation of iron- and arsenic sulfides). For the optimal performance of zero-valent iron filters, redox conditions have to be controlled, for example by adapting the filter designs to local water compositions.

This presentation will give a brief overview of the relevant processes. It shows that more work is needed to fully understand the important reaction pathways and to further optimize and expand the range of arsenic removal methods. One of the major challenges is the large variability in the concentrations of the most important water constituents that affect the performance and operation time of arsenic removal units.

SELECTIVE ION EXCHANGE FOR REMOVAL OF INORGANIC TRACE CONTAMINANTS FROM WATER

Wolfgang H. Höll; Forschungszentrum Karlsruhe

Introduction

Various natural and anthropogenic reasons lead to the presence of heavy metals in natural waters: (i) Mobilisation from soils or rocks during the natural water circulation, (ii) discharge of effluents of actual or former mining, (iii) discharge of waste effluents e.g. from the mineral processing and metal finishing industries, (iv) unlawful discharge of waste water into the underground by individuals, v) leaching of solid waste from landfill deposits or unlawful waste deposition, and (vi) oxidation of sulphides by nitrate input after applying fertilizers in agriculture.

Relevant metal contaminants are mercury, cadmium, arsenic, nickel, uranium, and chromium because of their toxicity. Corresponding with the local situation further metals causing undesirable effects (copper, cobalt, molybdenum, barium, selenium, antimony) may be encountered in concentrations of up to 100 µg/L, in exceptional cases even 2000 µg/L. Because of the well-known harmful effects of heavy metals on human health the concentrations of the inorganic species in drinking water should not exceed standards established by WHO or by individual countries (1).

The speciation of most of the metals strongly depends on the composition of the aqueous phase and may be comprised of free cations, inorganic complexes, complexes with natural organic substances or complexes with strong complexing agents. Among the metals eliminated in the respective studies, mercury can be present in a number of forms and, through interactions with the components of the water matrix, can be transformed into a variety of structures. The most commonly known forms include elemental mercury, divalent mercury and methyl mercury. The speciation of divalent transition metals includes cations Me^{2+} but also complexes with different ligands, depending on the background composition of the water. Chromium is usually present as monovalent oxide anions $HCrO_4^-$. Uranium forms stable complexes with carbonate ions and is present as e.g. $UO_2(CO_3)_2^{2-}$ carbonato complex species (2).

Basic principles of elimination

In raw waters of drinking water supplies heavy metals normally occur at trace concentration levels of up to several 100 µg/L at the maximum while the components of the background composition of natural waters have much higher concentrations. Removal of trace contaminants can be achieved by a variety of demineralisation methods, including membrane processes, chemical precipitation, and sorption processes. However, in most cases only the toxic contaminants have to be eliminated. As a consequence, a selective removal without affecting the background composition is desirable. Such a selective separation is not possible by means of membrane processes. Application of lime precipitation methods is limited by the solubility of hydroxides, which, in some cases, does not allow reaching the required standards. Furthermore, there is an often unwanted softening. A selective elimination of target contaminants in the trace concentration range is exclusively possible by means of ion exchangers. Among the suitable materials weakly basic resins offer particular advantages (3).

Elimination of heavy metal cations

Divalent heavy metal cations can be eliminated from natural waters by adsorption onto weak base anion exchangers. The adsorption is due to the presence of a nitrogen atom in the functional group. Its free electron pair leads to coordinative bonding of transition metals. Nitrogen atoms are also part of the functional groups of weakly basic anion exchangers. As a consequence, they are also able to adsorb heavy metal species. Unlike in cation exchangers, however, the charge of the metal cations is not compensated by negative charges of the functional groups. Therefore, to maintain electroneutrality, the adsorption of metal cations has to be accompanied by the simultaneous adsorption of an equivalent amount of anions. As a total, the process develops as the adsorption of heavy metal salts. Using cadmium and sulphate as an example, the service cycle can be written as (3):

$$\overline{R-NH_2} + Cd^{2+} + SO_4^{2-} \Leftrightarrow \overline{R-NH_2(Cd^{2+}, SO_4^{2-})} \quad (1)$$

R designates the resin matrix; overbarred symbols refer to the resin phase. The elimination of heavy metal salts other than $CdSO_4$ develops in a similar way.

Protons are strongly preferred over heavy metal species by the nitrogen atoms of the functional groups. With decreasing pH, therefore, the uptake of acids increases and the metal salts are removed. Consequently, desorption of heavy metals can be achieved by treatment with a sufficiently strong acid.

$$\overline{R-NH_2(Cd^{2+}, SO_4^{2-})} + H_2SO_4 \Leftrightarrow \overline{R-NH_3^+HSO_4^-} + CdSO_4 \quad (2)$$

Because of the strong preference of acids the exchanger cannot be reused in this form. In a second step of the entire regeneration it has to be reconverted to the free base form by means of caustic.

$$\overline{R-NH_3^+HSO_4^-} + 2\,NaOH \Leftrightarrow \overline{R-NH_2} + Na_2SO_4 + 2\,H_2O \quad (3)$$

The advantage of this kind of process lies in the fact that alkaline earth cations being the main constituents of natural waters do not undergo coordinative bonds with the nitrogen atoms. As a consequence, there is an extreme selectivity for heavy metal salts. The disadvantage of this process is that the rate of adsorption is generally small and cannot be increased.

Elimination of heavy metal anions

Selective removal of a spectrum of oxy anions is also possible by means of weakly basic anion exchangers. The reason is due to the protonation of weakly basic exchangers even a neutral pH and to the strong preference of e.g. chromate and also $UO_2(CO_3)_2^{2-}$ over sulphate [4]. Elimination of chromate, therefore, develops as:

$$\overline{R\text{–}(N(R')_2 \bullet H_2O)_2} + (CrO_4^{2-}) \Leftrightarrow \overline{R\text{–}(NH(R')_2^+)_2\,(CrO_4^{2-})} + 2\,OH^- \quad (4)$$

In natural waters the release of hydroxyl ions is buffered by the presence of carbonic acid. Regeneration is achieved by using sodium hydroxide solutions:

$$\overline{R\text{–}\,(N(R')_2^+)\,CrO_4^{2-}} + 2\,NaOH \Leftrightarrow \overline{R\text{–}(N(R')_2 \bullet H_2O)_2} + 2\,Na^+ + CrO_4^{2-} \quad (5)$$

Elimination of uranium species follows the same principle:

$$\overline{R\text{–}(N(R')_2)_2 \bullet H_2O} + UO_2(CO_3)_2^{2-} \Leftrightarrow \overline{R\text{–}(N(R')_2)_2^{2+}\,UO_2(CO_3)_2^{2-}} + 2\,OH^- \quad (6)$$

In the case of uranium removal regeneration has to follow a different way. Application of NaOH leads to a conversion of divalent to tetravalent uranium carbonato complex anions which are very strongly bound and which are only partly displaced (4). Regeneration has to start with a strong acid to convert the complex anions to UO_2^{2+} cations which cannot be sorbed by the anion exchanger acidic conditions.

$$\overline{R\text{–}(N=(R')_2)_2^{2+}\,UO_2(CO_3)_2^{2-}} + H_2SO_4 \Leftrightarrow \overline{R-(N=(R')_2)_2^{2+}\,SO_4^{2-}} + UO_2^{2+} + H_2CO_3 + CO_3^{2-} \quad (7)$$

In a second step NaOH is applied to convert the anion exchangers to the free base form:

$$\overline{R\text{–}N(R')_2^{+}\,SO_4^{2-}} + 2\,NaOH \Leftrightarrow \overline{R\text{–}N(R')_2 \bullet H_2O} + 2Na^{+} + SO_4^{2-}) \quad (8)$$

Materials and methods

Exchange resins

For the investigations only commercially available weakly basic anion exchangers have been applied. A manifold of resins were applied to find out the most effective once for the removal of each heavy metal. **Table 1** summarizes the exchangers used in the experiments that are reported (5, 6).

For the laboratory scale experiments the resins were subjected to a standard pre-treatment with hydrochloric acid and sodium hydroxide. In the pilot scale experiments the resins were used as delivered after intensive rinsing.

Table 1: Ion exchange resins applied in the experiments

Name	Manufacturer	Matrix	Functional Group
Duolite A7	Rohm and Haas	Phenol - Formaldehyde	Primary and secondary amine
Amberlite IRA67 / PWA 8	Rohm and Haas	Acrylic – DVB	Tertiary and secondary amine
Purolite A 845	Purolite Intern.	Acrylic – DVB	Tertiary and secondary amine

Column Installations

The laboratory scale experiments were carried out using filter columns of 2 3 cm inner diameter and bed heights of to 5-30 cm. Raw water from a reservoir was pumped across the columns in upstream direction. pH was measured and recorded and samples were taken from the effluent. In the laboratory scale experiments the raw water consisted of tap water from the Research Center spiked with the respective heavy metal. Throughput amounted to 5-20 bed volumes per hour (BV/h) during service cycle and to 2 – 2.5 BV/h during regeneration, which was carried out in counterflow. (7).

Pilot scale experiments for elimination of traces of mercury and chromate were carried out in the framework of a cooperation project with Tsinghua University, Beijing, China. The pilot plant had two columns of 25 cm diameter, each containing 75 L of exchanger material. Downstream throughput amounted to 10 - 20 BV/h. Regeneration was again carried out in counterflow (8).

Further pilot scale experiments for removal of traces of natural uranium were carried out in Germany. The respective plant contained two columns of 0.5 m diameter, each filled with 300 L of exchanger material. No regeneration was carried out.

Heavy metal concentrations in feed and effluent were determined by means of AAS or ICP-MS of samples obtained from automatic samplers.

Results and discussion

Removal of cations

Results from pilot scale experiments with respect to elimination of mercury species from contaminated lake water close to Haikou, Hainan Island, China are shown in

Figure 1. For this experiment the weakly basic exchanger Purolite A 845 was applied. The polluted feed water contained about 1.7 µg/L Hg. Up to a total throughput of about 16,000 BV the effluent mercury concentration was smaller than the drinking water standard of 1 µg/L. No regeneration could be carried out at the site (8).

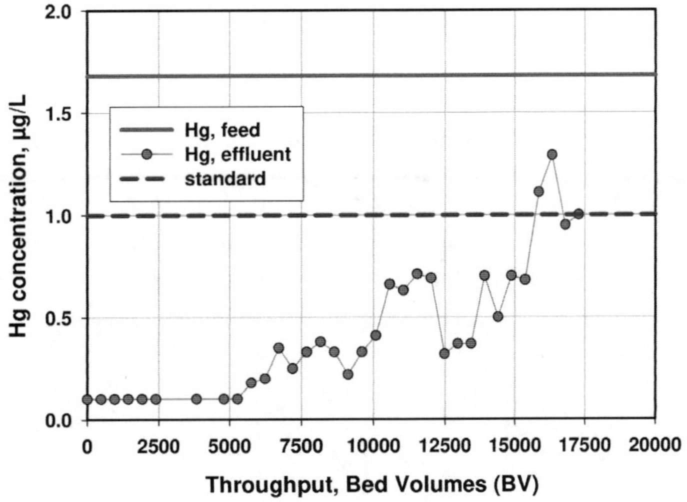

Figure 1: Development of mercury elimination at Haikou, Hainan, China. Resin: Purolite A 845, throughput: 20 BV/h.

Removal of oxy anions

Removal of oxy anions has been demonstrated in the pilot scale in the framework of the cooperation project with Tsinghua University, Beijing. Experiments were carried out at Shenyang, Northern China using a contaminated groundwater bearing 90 µg/L of hexavalent chromium. Purolite A 845 and Duolite A 7 were applied in the experiments. For the acrylic anion exchanger Purolite A 845 repeated service cycles of about 4,500 BV between two regenerations could be achieved. In constrast to this result, the development of the breakthrough curve for Duolite A7 reveals an unusually long service cycle (**Figure 2**). After 40,000 BV throughput the effluent Cr concentration was only 25 µg/L and still far below the drinking water standard of 50 µg/L (8). Analysis of the development showed that chromate oxidised the exchanger and was thus converted to chromium hydroxide.

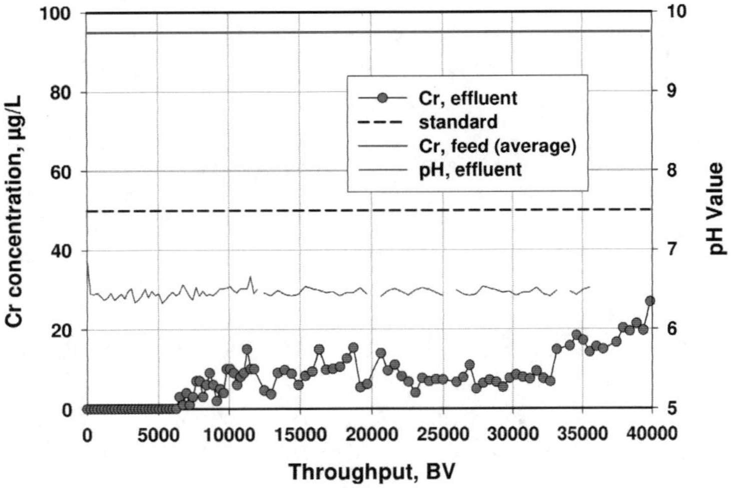

Figure 2: Development of Cr(VI) concentration in pilot scale experiments at Shenyang, China. Resin: Duolite A7, throughput: 20 BV/h.

Regeneration is possible; because of the oxidation, however, the exchanger is not re-usable. Despite this problem, the length of the service cycle seems to be long enough to use the exchanger in a one-time application.

Removal of natural uranium has been investigated in the laboratory and pilot scales. **Figure 3** shows results from pilot scale experiments in a municipal water work for a relatively small feed concentration of 10 – 13 µg/L. Up to t total throughput of 60,000 BV the effluent concentration was below 0.1 µg/L (left). Analysis of standard anions demonstrated that no change of the background composition occurs. Furthermore, no release of organic compounds or amines by the anion exchangers could be detected.

Regeneration of the uranium loaded exchangers has also been studied. By treating the resin with 0.5 molar H_2SO_4 in the first step and with 1 molar NaOH in a second step a nearly complete elution of the uranium was achieved.

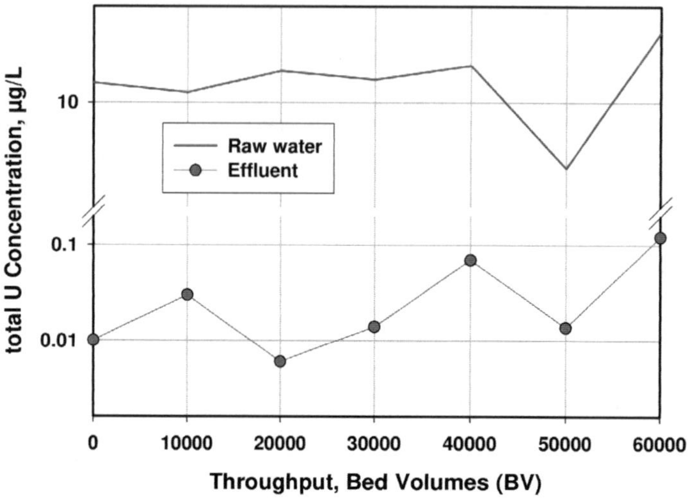

Figure 3: Development of uranium elimination in two pilot scale experiments. Resin: Amberlite IRA 67 / PWA 8, throughput: 20 BV/h.

Conclusions

The experiments have demonstrated that weakly basic anion exchangers in their free base form allow an efficient elimination of heavy metals, both of divalent cations and oxy anions. The successful removal of cadmium, mercury, nickel, uranium, and chromium has been demonstrated in the pilot scale with real raw waters. Depending on the raw water concentrations operation times of up to 10,000 BV can be obtained.

Acknowledgements

The authors thank the German Ministry of Research and Education (bmbf) and the German Association of Gas and Water (DVGW) to support this project

References

1. Hein, H., Schwedt, G. (1996). Richt- und Grenzwerte, Luft, Wasser, Boden, Abfall, Chemikalien, Vogel-Verlag Würzburg
2. Bernhard, G., Geipel, G., Brendler, V., Nitsche, H. (1998), „Uranium speciation in waters of different uranium mining areas“, J. Alloys and Compounds 271 – 273, 201 – 205.

3. Höll, W. H. (1996). "Elimination of heavy metals from water by means of weakly basic anion exchange resins". Proc. IEX '96, J. Greig (ed.), Soc. of Chem., Hartwells Ltd, Bodmin, UK, 404 – 411.
4. Riegel, M. Höll, W. H. (2008), "Uranium removal from natural waters using weakly basic anion exchangers", in: IEX 2008, Recent Advances in Ion Exchange Theory & Practice, Proc. of IEX2008, Cambridge, M. Cox, Editor, 331 – 338
5. Rohm and Haas Comp., Product information.
6. Purolite International, Product information.
7. Höll, W., Bartosch C., Zhao, X., He, S. (2002). "Elimination of trace heavy metals from drinking water by means of weakly basic anion exchangers", J. Water Supply: Research and Technology - AQUA 51 (3), 165 - 172.
8. Höll, W. H., Zhao, X., Hagen, K. (2004), "Elimination of health-relevant heavy metals from raw waters of the drinking water supply in the PO China by means of weakly basic anion exchange resins, Report FZKA 6994, Forschungszentrum Karlsruhe.

91d Toxic Heavy Metal Removal by Hollow Fiber Membrane Solvent Extraction-Based Techniques

Kamalesh K. Sirkar, Chemical Engineering, New Jersey Institute of Technology, Membrane Seperations, 161 Warren st, Newark, NJ 07102

The nondispersive membrane solvent extraction technique was developed in the early eighties.It is now being used in a number of large-scale commercial installations. In the very late eighties this technique was explored for heavy metal extractions as well. The nineties saw a very large number of laboratory scale studies and a number of pilot scale studies continuing into the early part of this decade. This talk will illustrate this technique for extraction as well as recovery of various toxic heavy metals. The strengths and weaknesses of this technique will be pointed out. Other related techniques developed will be illustrated as well along with a few larger-scale exploitations.

91e Endocrine Disrupting Chemical Risk Management Research in the US Epa's Office of Research and Development

Carolyn Acheson[1], Marc Mills[1], Steve Hutchins[2], Todd Martin[1], Gregory Sayles[1], Kathleen Schenk[1], and Douglas Young[1]. (1) National Risk Management Research Laboratory, US EPA, MS 421, 26 W Martin Luther King Dr, Cincinnati, OH 45268, (2) Ground Water and Ecosystem Restoration Division (GWERD), National Risk Management Research Laboratory, US EPA, P.O Box 1198, Ada, OK 74821-1198

Endocrine disrupting chemicals (EDCs) are receiving increasing media and scientific attention. Concerns about these chemicals stem from the possibility of serious human and wildlife effects and environmental persistence. The US EPA Office of Research and Development's National Risk Management Research Laboratory is conducting research to investigate sources of suspected EDCs that impact the environment and to evaluate risk management strategies to minimize exposure of humans and wildlife to suspected EDCs. In the EDC research program, risk management research is occurring concurrent with health effects, exposure, and risk assessment research.

Early in the research program, a risk management evaluation was used to identify chemicals which were likely to be endocrine active and environmental sources of these compounds. Based on this evaluation, risk management EDC research efforts focused on activities such as wastewater treatment, concentrated animal feeding operations (CAFOs), drinking water treatment, and pollution prevention. Chemicals of concern include steroid hormones and alkyl phenol ethoxylates; other chemicals may be added as effects, exposure and assessment research identifies concerns.

To date, research efforts have focused on determining the capability of conventional wastewater, CAFO effluent, and drinking water treatment technologies to remove EDCs. As effects, exposure and assessment research identifies unacceptable risks, future research may evaluate new or improved methods to cost effectively manage the risk. Pollution prevention research has evaluated chemical substitutes for endocrine active compounds such as alkyl phenol ethoxylates. An overview of these projects will be presented.

168a Selective and Regenerable Ion Exchange for Perchlorate Removal, Recovery, and Environmental Forensics

Bachua Gu, Environmental Sciences Division, Oak Ridge National Laboratory, Oak Ridge, TN 37381

Selective and regenerable ion-exchange technology has been developed for removing perchlorate (ClO_4^-) from contaminated water at trace concentrations. The technology relies on a unique, bifunctional anion-exchange resin to sorb ClO_4^- selectively, and the resin is then regenerated using tetrachloroferrate displacement techniques. The regeneration technology enables quantitative recovery of perchlorate from contaminated water or other environmental matrices such as plants, sediments and/or salt deposits for possible reuse or for isotopic source identification. Naturally-forming perchlorate has been found to contain distinct oxygen and chlorine isotope signatures as compared with anthropogenic perchlorate, thereby allowing unambiguous identification of sources of perchlorate for environmental forensics.

168b Adsorption of Trace Levels of Arsenic from Aqueous Solutions by Conditioned Layered Double Hydroxides: Flow Experiments and Multi Solute Batch Experiments

Megha Dadwhal, Muhammad Sahimi, and Theodore T. Tsotsis. Mork Family Department of Chemical Engineering and Materials Science, University of Southern California, 925 Bloomwalk Avenue, HED 216, Los Angeles, CA 90089

Arsenic is found in water in the form of oxyanions. Relatively high concentrations of As have been reported both in power plant discharges, as well as, in fresh water supplies. Inorganic As may be present in the 3+ (As(III), arsenite) and 5+ (As(V), arsenate) oxidation states[1]. Protracted contact with As-containing water is thought to cause arsenicosis, a form of poisoning in humans[2]. The International Agency for Research on Cancer (IARC) currently classifies As as a group 1 chemical, which is considered to be carcinogenic to humans.

The focus of the present work was to study the removal of As(V) by conditioned, calcined layered double hydroxide (LDH) adsorbents in continuous-flow, packed-bed columns in order to study the effect of important operating parameters, such as the influent As concentration, the pH, the adsorbent particle size, and the flow rate. As expected, an earlier bed saturation and breakthrough was observed at higher flow rates and influent concentrations. On the other hand, a decrease in the adsorbent particle size and the influent pH resulted in an increase in the number of bed volumes at breakthrough. A column model which accounts for external, liquid-film mass transport and for diffusion and adsorption in the adsorbent particles is utilized. Two different adsorbent particle models are employed, which were shown previously capable of predicting As(V) uptake by LDH adsorbents. A conventional homogeneous surface diffusion model, and a bidisperse pore model; the latter model views the LDH particles as assemblages of microparticles and takes into account bulk diffusion in the intraparticle pore space, and surface diffusion within the microparticles themselves. Both models are found capable of predicting the flow-column experimental results.

Batch experiments with simulated power plant effluents were also carried out. The effect of the presence of anions, namely NO_3^{2-}, SO_4^{2-}, PO_4^{2-}, CO_3^{2-}, Cl^- and F^-, on arsenic adsorption was studied. We collected data on various power plant effluents in the greater Los Angeles area, in order to identify the concentrations of these ions present in the power plant effluents. LDH was found to be effective adsorbent to remove arsenic in the presence of various ions in the solution.

References:

(1) Ferguson, J.F.; Gavis, J.; A Review of the Arsenic Cycle in Natural Waters. Water Res. 1972, 6, 1259.

(2) Lepkowski, W.; Arsenic Crisis in Bangladesh. Chem. Eng. News 1998, 76 (46), 27.

(3) Yang, L.; Shahrivari, Z.; Liu P.K.T.; Sahimi, M.; Tsotsis, T.T. Removal of Trace Levels of Arsenic and Selenium from Aqueous Solutions by Calcined and Uncalcined Layered Double Hydroxides (LDH), Ind. Eng. Chem. Eng. 2005, 44(17), 6804-6815.

(4) Yang, L; Dadwhal, M; Shahrivari, Z; Ostwal, M; Liu, P.K.T; Sahimi, M; Tsotsis, T.T. Adsorption of Arsenic on Layered Double Hydroxides: Effect of the Particle Size. Industrial & Engineering Chemistry Research , 2006, 45(13), 4742-4751.

Keywords: Layered double hydroxides, Homogeneous surface diffusion model, Bidisperse pore model, Arsenic, Adsorption, Packed bed columns, Multi solute.

168c Remove Arsenic and MS2 Virus by Iron Oxide Nanoparticles Coated on Glass Fibers

Xuan Li[1], Jinwen Wang[1], Leonardo A. Gutierrez[2], and Thanh H. Nguyen[2]. (1) Department of Materials Science & Engineering, University of Illinois at Urbana-Champaign, 1304 W Green Street, Urbana, IL 61801, (2) Department of Civil and Environmental Engineering, University of Illinois at Urbana-Champaign, 205 N Mathews Avenue. Room 4146, Urbana, IL 61801

Arsenic poisoning and viruses are two of the most widespread water-related problems in the world. A new generation of adsorption and ion exchange materials has been developed in our laboratory. These fibers offer orders of magnitude increase in capacity and kinetics over conventional technology. Additionally, the fibers can be manufactured from a number of precursor materials, formed on a variety of substrates due to a unique low temperature process, and can be tailored for high adsorption capacity, adsorption kinetics and selectivity. We have optimized and scaled up a low-cost system consisting of iron oxide supported on glass fiber substrates which effectively removes the As3+ and As5+ from groundwater (below 1ppb). Similarly, MS2 virus was indicated to well below the USEPA standards.

Chromium (III) Sequestration by Oats Straw and Agave Baggasse: Sorption Mechanism

Refugio Bernardo García-Reyes and José René Rangel-Méndez
Instituto Potosino de Investigación Científica y Tecnológica, A.C., San Luís Potosí, S.L.P, México.

Abstract

Agro-waste materials such as oats straw (OS) and agave baggasse (AB) have functional groups which are able to bind metal ions. The agro-waste materials studied in this research (acid washed) were chemically characterized and their chromium sorption and desorption capacity was obtained. HNO_3, NaOH and EDTA (at different concentrations and temperatures) were used to regenerate partially saturated biosorbents. Finally, the chromium (III) sorption mechanism was discussed.

The results demonstrated that OS and AB contain carboxyl and hydroxyl functional groups. The chromium (III) removed by acid washed OS and AB was 2.48 and 9.85 mg/g, respectively. Based on chromium (III) removed and ions released (mainly H^+ and Ca^{2+}) during the sorption process, it was suggested that the main sorption mechanism was ion exchange: for each mol of $CrOH^{2+}$ adsorbed 1 mol of Ca^{2+} and 2 mol of H^+ were released to the solution. On the other hand, the highest chromium desorption was obtained by using an EDTA solution at 55°C.

Oats straw and agave baggasse are capable of removing cations from aqueous solution. Preliminary results show that chromium (III), as $CrOH^{2+}$, is exchanged by Ca^{2+} and H^+ during the sorption process. In addition, an EDTA solution could be used to regenerate the chromium loaded biosorbents.

Keywords: Oats straw, Agave baggasse, Chromium (III), Sorption, Mechanism.

Introduction

Wastewater containing chromium is discharged from industries such as metallurgical, electronic, tannery, metal plating, among others. It is well known that hexavalent chromium is extremely toxic and carcinogenic. However, hexavalent chromium could be reduced to trivalent chromium by some microorganism or manganese oxide. Recently, Suwalsky *et al.* (2008) demonstrated that trivalent chromium exerts stronger structural effects than hexavalent chromium on the human erythrocyte membrane and membrane model. These new findings lead us to look for alternatives to remove chromium (III) from aqueous solution.

Biosorption is an emerging technology for treating water contaminated with heavy metal ions. For instance, seaweed biomass (Yun *et al.*, 2001; Cossich *et al.*, 2004), and agricultural by-products (Romero-Gonzalez *et al.*, 2006; Sawalha *et al.*, 2007 and 2008) have been used for removing chromium (III) from water.

Agro-waste materials components (cellulose, hemi-cellulose, pectin and lignin) contain carboxyl and hydroxyl functional groups which can be involved in metal binding (Reddad *et al.*, 2002). However, desorption studies are not often reported when agro-waste materials are used as adsorbents. In addition, the sorption mechanism is not well understood.

For that reason, the objectives of this research are to chemically characterize agro-waste materials (oats *sativa* straw (OS) and agave *salmiana* baggasse (AB)) locally available in Mexico, and to obtain the chromium (III) sorption/desorption capacity. Finally, the chromium (III) sorption mechanism was discussed based on results previously reported in literature and the ions released during the adsorption process.

Methodology

Agricultural residues such as oats *sativa* straw (OS) and agave *salmiana* baggasse (AB) were used as adsorbents after being hydrochloric acid (0.01 N) washed (A). These capital letters were used throughout the document to identify each biosorbent; for example acid washed agave *salmiana* baggasse is represented as AAB. To obtain the agro-waste materials surface charge distribution, potentiometric titrations were carried out and to estimate the functional groups and equilibrium constants, the experimental data obtained from potentiometric titration were processed according to the method proposed by Yun *et al.* (2001). Attenuated Total Reflection Fourier Transform Infrared (ATR-FTIR) was used to identify the biosorbents functional groups.

Batch sorption experiments were carried out to determine the chromium (III) adsorption capacity. Then regeneration of partially saturated agro-waste materials was carried out by using 100 mL of HNO_3 (0.1 and 1.0 N), NaOH (0.1 and 1.0 N) or EDTA (0.05 and 0.1 M) at 25, 35 and, in some cases, 55°C. The initial and equilibrium chromium and calcium concentration were determined by an atomic absorption spectrophotometer.

Results and discussion

The functional groups present in AOS and AAB were similar, although these exist in different amounts. The first functional group can be associated to carboxyl (pk_a value 3.8-4.1), these were confirmed by ATR-FTIR analyses at 1745, 1650, 1260, 1160, and 1060 cm^{-1} (Wade, 1993; Socrates, 1998). The second functional group cannot be associated to a specific group by the great amount of functional groups (amine, phosphoryl, amide, imidazol, etc.) associated to the pk_a values between 6.3 and 7.1. Finally, the third functional group can be associated to hydroxyl groups (pk_a 9.1-12.6) and these groups were confirmed by ATR-FTIR analysis in a broad band at 3350 cm^{-1}.

The pH_{PZC} was 3.2 and 3.5 for AOS and AAB, respectively. Because the adsorption experiments were carried out at pH 4, the agro-waste materials net surface charge posses a negative charge and, for that reason, cations metal could be adsorbed by electrostatic attraction. For instance, chromium species are positively charged at pH 4 as follows: Cr^{3+} (30%), $CrOH^{2+}$ (40%), $Cr_2(OH)_2^{4+}$ (26%) and $Cr_3(OH)_4^{5+}$ (4%). These species could be adsorbed in negatively charged functional sites (e.g. carboxyl groups).

The amount of chromium adsorbed and calcium released is showed in Figure 1. AOS adsorbed less chromium and released less calcium than AAB. These findings demonstrated that AOS contain less calcium than AAB, and these calcium released is directly related to the amount of chromium adsorbed. For example, for each mol of chromium (III) adsorbed 0.1 and 1.2 mol of calcium were released from AOS and AAB, respectively.

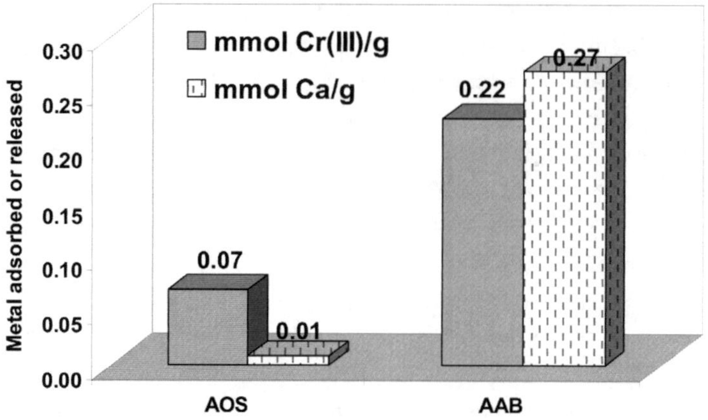

Figure 1. Chromium adsorbed and calcium released in agro-waste materials at pH 4 and 25°C.

However, the chromium adsorbed in AOS can not be explained by the release of calcium. For that reason, an additional adsorption experiment was carried out at the same conditions (25°C, 100 mg of AOS in 100 mL) but at pH without control (starting at pH 4) and initial chromium concentration of 100 mg/L. The initial and final pH was measured and then the amount of H^+ and chromium (III) ions was calculated. The molar ratio average, between H^+ released/chromium adsorbed, was 1.93 which suggested that $CrOH_2^+$ is preferentially adsorbed onto the agro-waste materials because when this specie is adsorbed, 2 mol of H^+ are released to the solution according to the following reaction (1):

$$2\{\equiv R\text{-}COOH\} + CrOH^{2+} \leftrightarrow \{\equiv R\text{-}COO\}_2\text{–}CrOH + 2H^+ \qquad (1)$$

In contrast, based on chromium (III) removed and calcium released during the sorption process (Figure 1), the adsorption mechanism could be:

$$\{\equiv R\text{-}COO\}_2\text{-}Ca + CrOH^{2+} \leftrightarrow \{\equiv R\text{-}COO\}_2\text{–}CrOH + Ca^{2+}$$

On the other hand, although the agro-waste materials presented less sorption capacity for chromium than seaweed biomass, activated carbons, or ion exchange resins, the agro-waste adsorbents are an available and cost-effective option for the removal of chromium (III) from aqueous solution. However, it is necessary to find a way to regenerate them and reuse them to make their application more attractive in real systems. In that sense, batch desorption experiments were carried out by using chromium partially saturated biosorbents. The chromium sorption capacity, at pH 4 and 25°C, was 9.9±1.0 and 2.5±1.0 mg/g for AAB and AOS, respectively. Chromium desorption results by HNO_3, NaOH and EDTA are given in Figure 2.

From Figure 2, it is clearly observed that the agro-waste materials regeneration was affected by both temperature and the eluent concentration. For instance, the regeneration by EDTA 1.0 M at 25 and 35°C was 32 and 39%, respectively, but at 55°C it achieved 63%. In addition, regeneration by EDTA did not cause initial weight lost as 1.0 N HNO_3 or NaOH (results are not shown).

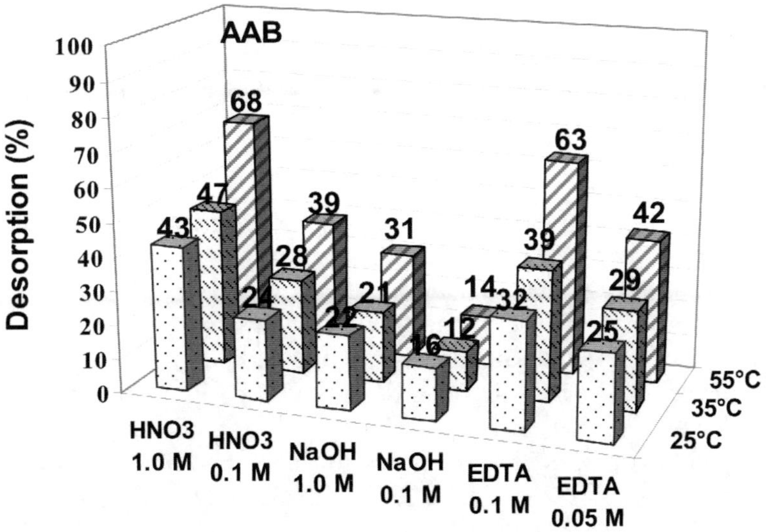

Figure 2. Chromium (III) desorption from AAB with different eluents and temperatures.

However, the low percentage of chromium desorbed by HNO_3 (1.0 N) at 25°C (see Figure 2) indicated that chromium (III) ions were covalently bound to the biosorbents functional groups (e.g. oxygen-containing sites). Finally, chromium desorption from agro-waste materials by EDTA at 55° presented an interesting option.

Conclusion

Oats straw and agave baggasse contain carboxyl and hydroxyl functional groups which are able to bind chromium (III) from aqueous solution. Preliminary results show that chromium (III), as $CrOH^{2+}$, is exchanged by Ca^{2+} and H^+ during the sorption process. In addition, an EDTA solution can be used to regenerate 63% of the chromium-loaded biosorbents without apparent modifications on the adsorbent.

Acknowledgements

R.B. García-Reyes would like to thank CONACyT for the scholarship (204219) received. Also, the authors thank the Environmental Science Division of IPICyT, for the infrastructure and Dulce Partida for the technical support to carry out the project.

References

Cossich ES, Da Silva EA, Tavares CRG, Cardozo L, Ravagnani TMK (2004), Biosorption of chromium(III) by biomass of seaweed *Sargassum sp.* in a Fixed-Bed Column, Adsorption, **10(2)**, 129–138.

Reddad Z, Gerente C, Andres Y, Le Cloirec P (2002), Adsorption of several metal ions onto a low-cost biosorbent: Kinetic and equilibrium studies, Environmetal Science Technology, **36(9)**, 2067-2073.

Sawalha MF, Peralta-Videa JR, Saupe GB, Dokken KM, Gardea-Torresdey JL (2007), Using FTIR to corroborate the identity of functional groups involved in the binding of Cd and Cr to saltbush (Atriplex canescens) biomass, Chemosphere, **66(8)**, 1424–1430.

Sawalha MF, Peralta-Videa JR, Duarte-Gardea M, Gardea-Torresdey JL (2008), Removal of copper, lead, and zinc from contaminated water by saltbush biomass: Analysis of the optimum binding, stripping, and binding mechanism, Bioresource Technology, **99(10)**, 4438–4444.

Socrates G. (1998) Infrared Characteristic Group Frequencies, 2nd ed., John Wiley & Sons, New York, 249 pp.

Suwalsky M, Castro R, Villena F, Sotomayor CP (2008), Cr(III) exerts stronger structural effects than Cr(VI) on the human erythrocyte membrane and molecular models, Journal of Inorganic Biochemistry, **102(4)**, 842–849.

Wade, L.G. Jr. (1993), QUIMICA ORGANICA, 2nd. Ed., Prentice Hall, 1312 pp.

Yun YS, Park D, Park JM, Volesky B (2001), Biosorption of trivalent chromium on the brown seaweed biomass, Environmental Science Technology, **35(21)**, 4353-4358.

Romero-Gonzalez J., Peralta-Videa J.R., Rodrıguez E., Delgado M., Gardea-Torresdey J.L. (2006), Potential of agave *lechuguilla* biomass for Cr(III) removal from aqueous solutions: thermodynamic studies, Bioresource Technology, **97(1)**, 178–182.

168e Hybrid Ion Exchanger (HIX): Characterization and New Application Opportunities

Arup K. Sengupta, Department of Civil & Environmental Engineering, Lehigh University, 13 E. Packer Avenue, Bethlehem, PA 18015

Hybrid ion exchangers (HIXs) are essentially inorganic or polymeric ion exchangers within which an external agent, namely, ion, solvent, nanocrystal, nanoprecipitate or a combination of the above have been deliberately introduced to develop unique properties, not otherwise attainable. Previously, hybrid ion exchangers were synthesized with paramagnetic properties and high sorption affinity toward arsenic oxyanions or oxyacids. Here we report both unique sorption behaviors and toxic metal sensing ability of a new class of polymeric/inorganic hybrid ion exchangers.

227a Arsenic Removal from Drinking Water Using a Regenerable Adsorbent

Paul Sylvester and Teresia Moller. SolmeteX - A Division of Layne Christensen, 50 Bearfoot Road, Suite # 2, Northborough, MA 01532

Introduction

The consumption of water containing low levels of arsenic has been linked to an increased risk of lung cancer, bladder cancer and other human ailments. In January 2006, this risk prompted the United States Environmental Protection Agency (EPA) to reduce the maximum contamination level for arsenic in drinking water from 50 μg/L to 10 μg/L, bringing the United States in line with the European Union and the recommended World Health Organization (WHO) level. Naturally occurring arsenic is found in the groundwater in many parts of the United States and there are also areas where extended pesticide use has led to the anthropogenic contamination of groundwaters. The EPA has estimated that approximately 4,000 water providers and several million domestic wells were not in compliance with this new lower arsenic MCL

There are a number of methods to remove arsenic from groundwater that have successfully been applied on a commercial scale, including ion exchange, adsorption and coagulation-filtration technologies. Of these methods, probably adsorption has proven the most popular due to simplicity of operation, low levels of waste residuals and cost of treatment. The most common adsorbents commercially available consist of oxides of iron, titanium, aluminum and zirconium, though the use of aluminum is decreasing following the development of new media with greater arsenic capacity and better pH tolerance. Granular media are used in packed beds and arsenic-contaminated water is passed through, either up or down flow, arsenic is adsorbed onto the media and treated water sent through to the distribution system. The advantage of adsorbent media over ion exchange is that common anions in water, such as bicarbonate, sulfate and chloride, are not adsorbed by the media so the taste and other aesthetical qualities are not impacted. The operational life of adsorptive media is long and can vary from tens to hundreds of thousands of bed volumes (BVs) of treated water which can equate to months or years of use.

When arsenic breakthrough into the treated water is detected, the adsorbent media must be replaced with fresh media. Although the arsenic can readily be stripped from most of the adsorbents using strong caustic, the low physical integrity of the granules means that the regeneration process will cause significant media attrition leading to the generation of fines and an overall reduction in particle size. Consequently, the regeneration of most adsorptive media is not considered to be economically viable. Currently, spent media are disposed of once breakthrough occurs, leading to the generation of large volumes of arsenic-tainted solid waste. Although this waste passes the current Toxicity Characteristic Leaching Procedure (TCLP) and can therefore be disposed of in a non-hazardous waste landfill, recent research is suggesting that arsenic may be remobilized under landfill conditions.

To overcome the limitations of granular adsorbents, SolmeteX in collaboration with Lehigh University developed ArsenX. This media consists of hydrous iron oxide particles immobilized throughout a strong base anion exchange resin. The media is manufactured in the mixed sulfate/bicarbonate form and removes arsenic by adsorption; the ion exchange component serving mostly as a rigid support while the removal is performed by the iron oxide. Due to the improved physically integrity supplied by the polymer bead, this media can be efficiently regenerated using warm caustic solution and reused, thus minimizing the generation of solid waste. Field and laboratory studies have shown that minimal arsenic capacity is lost during the regeneration process making regeneration and reuse and attractive economic option.

Field Study

The performance of ArsenX in the field was evaluated at Paxton, MA using a 3 liter (0.79 gallons) bed of media. The media was packed into a two inch diameter column giving a column height of approximately 5 feet. The water chemistry at this site is shown below in Table 1. All of the arsenic was present as As(V) (arsenate).

Parameter	**Value**	**Units**
pH	6.7	
Arsenic	21	ppb
Iron	0.03	ppm
Hardness	89	ppm $CaCO_3$
Silica	13	ppm
Chloride	55.3	ppm
Sulfate	20.7	ppm
Fluoride	0.2	ppm
Phosphate	<150	ppb

Table 1. Water Chemistry at Paxton, MA

Water was passed down through the resin at a flow rate of 1 liter per minute, giving an empty bed contact time (EBCT) of 3 minutes. Sampling points were located along the length of the media column to allow the arsenic breakthrough to be monitored as it progressed down the column, allowing data to be gathered on the kinetics of arsenic adsorption. At the conclusion of the trial, no arsenic was detected in the column effluent and it was still below the 10 ppb MCL at sampling point S-4, equivalent to an EBCT of 2.2 minutes. The breakthrough curves for arsenic at S-4 and S-2, equivalent to an EBCT of just 0.73 minutes is shown below in Figure 1. Both curves overlap each other indicating that the arsenic uptake kinetics for ArsenX are extremely rapid.

Figure 1. Breakthrough Curves for ArsenX

A full scale arsenic removal system is currently being installed at this site and consists of 2 separate 100 gpm skid-based systems connected in parallel. Each individual skid consists of 2 resin tanks in series, each tank containing 29 cubic feet (217 gallons of media). This system is designed to allow each skid to operate on one tank only to allow spent media to be regenerated off site without interrupting the well operation. Based upon the pilot data shown in Figure 1, it is estimated that each column of ArsenX will treat in excess of 100,000 BVs of water before requiring regeneration. The treatment system is expected to become operational in late May of 2008.

227b Investigation of Extraordinarily High Cr(VI) Removal Capacity by a Weak-Base Anion Exchange Resin: The Mechanism of Reactive Ion Exchange

Sudipta Sarkar[1], Nicole Blute[2], Jin-Cheng Lin[1], and Arup K. SenGupta[3]. (1) Department of Civil and Environmental Engineering, Lehigh University, 13 E. Packer Avenue, Bethlehem, PA 18015, (2) Malcolm Pirnie, Inc, 1821 Wilshire Blvd, Suite 302, Santa Monica, CA 90403, (3) Department of Civil & Environmental Engineering, Lehigh University, 13 E. Packer Avenue, Bethlehem, PA 18015

In a recent pilot-scale study with contaminated groundwater at Glendale, CA, two weak-base anion exchange (WBA) (Duolite A-7 and SIR-700) resins exhibited at least one order of magnitude higher capacity than strong base anion exchange resins for Cr (VI) removal in presence of significantly high concentrations of competing anions like chloride, sulfate, etc. Comparison with earlier studies suggested, in addition to anion exchange, presence of other mechanism(s) is responsible for such high chromate removal capacity. It was also observed that there was a significant simultaneous removal of copper that was present in the groundwater along with chromate ions. Such a simultaneous removal of copper and chromate ions by the WBA resin also indicated a possible presence of polymeric ligand exchange phenomenon along with ion exchange. A series of laboratory investigations involving fixed bed column runs and other characterization studies were undertaken to find out the exact mechanism behind such a high chromate removal capacity. The results showed that there was an increase in pH associated with chromium removal by the resin. Also, the WBA resins pre-loaded with copper ions did not show any preference towards ligands like arsenates and phosphates over other common anions, thereby negating possibility of any polymeric ligand exchange. Further studies on exhausted WBA resin showed presence of Cr(III) species inside the ion exchanger, indicating reduction of influent Cr(VI) ions to Cr(III). Other ion exchange column runs and comparative FTIR studies of the virgin, exhausted and oxidized resins confirmed a structural change in the organic polymer phase of the resins. The results of the study revealed a new mechanism of reactive ion exchange. First, chromate ions are removed from the solution through anion exchange at the functional groups of the resin. Subsequently, chromate ions are reduced to form insoluble Cr(III) species within the resin beads. The redox reaction results in a renewal of the positively charged functional groups of the resin for further ion exchange and subsequent redox reaction. Redox reaction is a slower process than ion exchange and therefore, is the rate limiting step in the removal process. Interruptions in the column runs proved the presence of a rate limiting process with characteristics different from the usual intra-particle diffusion process. The presentation shall discuss the findings of multi-faceted studies carried out to investigate the mechanism of reactive ion exchange leading to high capacity of the WBA resins for chromate removal.

Low Cost - Sustainable - High Capacity Potable Water Treatment Media for Arsenic Removal

Rajiv Banavali, *Jose Antonio Trejo* ***and*** *Garth R. Parker*
Rohm and Haas Company, Spring House, PA USA

Introduction

According to the World Health Organization (WHO), arsenic (As) contaminated drinking water affects tens of millions of people in more than twenty countries. Arsenic poisoning is most prevalent in poorest villages in South Asia, particularly Bangladesh, eastern India, Nepal, and Vietnam. Traditional village life is based on farming in these areas and depends on monsoons. The social fabric is held together through strong family ties and religion. Extended families live in close proximity, often with half a dozen or more people in houses made of clay, locally produced bricks, or corrugated tin. In most villages, no electricity is available. Water is usually supplied by a hand pump in the backyard or near-by tube wells. Many of these wells deliver water which exceeds the WHO arsenic limit for potable water by a factor of 10–100. At least half of Bangladesh's 15 million tube wells exceed the levels considered safe.

Under such dire circumstances, a treatment to remove arsenic must be very low cost, convenient for women and children to use, and not subject to the introduction of pathogens. In all likelihood, a device or system should be placed directly in the home and should be easily available whenever water is needed for drinking or cooking. Additionally, the removal and safe disposal of arsenic-contaminated wastes which result from the treatment is also needed. The design criteria, shown in Figure 1, were to achieve low cost, use of indigenous materials, high performance, ease of use, non-hazardous residuals, and relatively safe disposal.

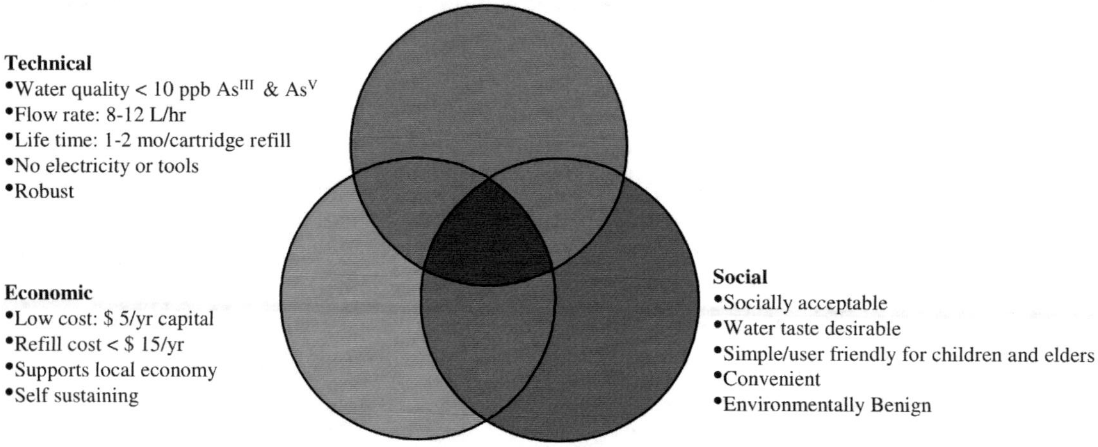

Figure 1: Sustainable chemistry cycle concept.

Novel Media

We have developed a novel media and potable water treatment system which meets our design criteria. This media can be used at the household scale, with design adaptability for larger scale community use, to remove arsenic from contaminated groundwater. The technology uses a modified limestone as the media support with an iron oxide coating. This media is unique because it efficiently removes arsenic and several other metal contaminants using a locally available raw material. The resulting drinking water quality complies with drinking water standards. The media, when fully loaded with arsenic, is considered a non-hazardous waste and can be easily disposed. The spent media when incorporated into concrete passes the US Environmental Protection Agency Toxicity Characteristic Leaching Procedure (TCLP) and California Wet tests.

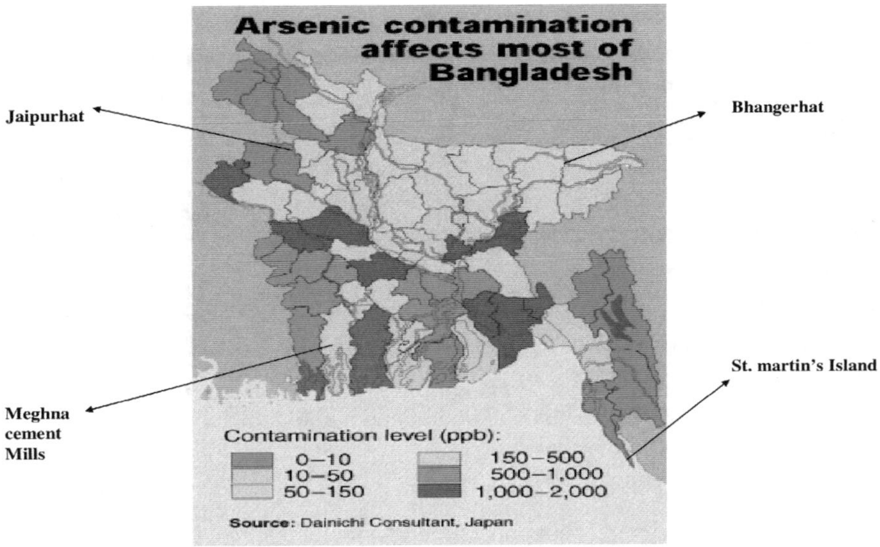

Figure 2: Arsenic Contamination and Major Limestone Deposits in Bangladesh

The map in Figure 2 above shows both the arsenic problem and availability of limestone. Bangladesh has deposits of limestone that are used to produce cement. Limestone deposits from coral and the shells of marine organisms are found near the land surface at St. Martin's Island at the southernmost tip of the country. Limestone also is found at Takerghat, in the northeastern part of the country, where it is mined in a quarry and in the subsurface at Joypurhat. The Jaipurhat Limestone Mining and Cement Works extract more than one million tons per year of limestone to operate its cement plant. Two other cement plants use about 1.1 million tons per year of limestone. In addition, a recent 25 km conveyer belt has been constructed to move limestone from nearby India to a large cement plant in Bangladesh, at the rate of 960 metric tons/hour. Even if all households in Bangladesh used our media (under the conditions as stated), only a fraction of the limestone Bangladesh currently uses annually would be required. Moreover, since much of the waste media could be recycled, there would be virtually no net loss of imported limestone. Our media consists of iron oxides limestone mixtures with the right balance to achieve high capacity for arsenic removal, less kinetic limitations than the commercially available iron oxides and a pH balance delivered by the limestone with good buffering capacity.

This unique combination not only permits the media to remove arsenic selectively from water but also other toxic metals which are removed by the pH effect and Ca delivered by the limestone material. Metals such as Hg, Pb, Cd, Fe are removed simultaneously with the arsenic. It is also important to mention that the removal of As(V) and As(III) by this media has been also demonstrated in field trials with great success. The arsenic removal efficiency of native limestone is virtually independent of pH values (ranging from 4 to10), making it highly suitable as a base media for a wide variety of water quality. No other technology can match this pH range adaptability. All of the materials are locally and regionally available or they will be manufactured directly in Bangladesh or India, thus contributing to the self-reliance and economy of the region. The components are highly cost-effective for this application.

Experimental

Materials: Iron oxide coated limestone (LS-FE) prepared in our laboratory, commercially available Granular Ferric Hydroxide (GFH).

Laboratory test solution: NSF-Challenge water with 100 µg/L As(V), 20 mg/L SiO_2, 140 µg/L PO_4^{3-}, adjusted to pH 7.5. The solution was stored in a 125 L polypropylene tank.

Laboratory test conditions: 1 cm ID glass column filled with 8 mL of media, giving a bed height of 10 cm. The solution was pumped through the column at a flow rate of 5 mL/min, giving a 1.6 min Empty Bed Contact Time (EBCT). Samples of effluent were taken at least every 6 hr using a fraction collector and tested for As using ICP-OES. Our breakthrough criteria is an effluent As concentration of 10 µg/L.

Field study test conditions: A system similar to that depicted in Figure 4 was installed in the home of a family in India at 24 South Parganas District located south of Kolkata. Raw water was obtained from the home's backyard well with a hand pump and transferred to a storage reservoir. Water flows by gravity through a media cartridge located under the storage reservoir and then up to a spigot for consumption. The cartridge used in the trial was 4 inch diameter PVC pipe packed with 0.5 L of untreated limestone and 1.5 L of LS-Fe. The gravity fed flow rate was approximately 250 mL/min (8 min EBCT). The flow was intermittent to simulate real use conditions. Samples were taken and measured for As using a Hach Field Test Kit and were sent to a local laboratory in Kolkata for quantitative results for As and other metals.

Laboratory Results

A comparison under laboratory conditions was done with two different materials: our novel iron hydrous oxide coated limestone (LS-FE) and GFH. Because the media is envisioned for a point of use application, high flow rate, short EBCT conditions were used in this study. Laboratory column testing of the LS-FE demonstrated improved capacity for As removal relative to standard media such as GFH, as indicated in Figure 3 and Table 1, where LS-FE treated 8× as much water. The capacity at high flow rate of the LS-FEe media is 1.7 g-As/L of media. LS-Fe was superior to GFH when high flow conditions are required.

Table 1. Bed volumes to Arsenic Breakthrough GFH and Iron Coated Limestone. Influent 100 µg/L As(V), EBCT = 1.6 min.

Media	Breakthrough Volume (BV)	Breakthrough Capacity (g As/L media)
GFH	1880	0.24
Iron coated limestone	15600	1.72

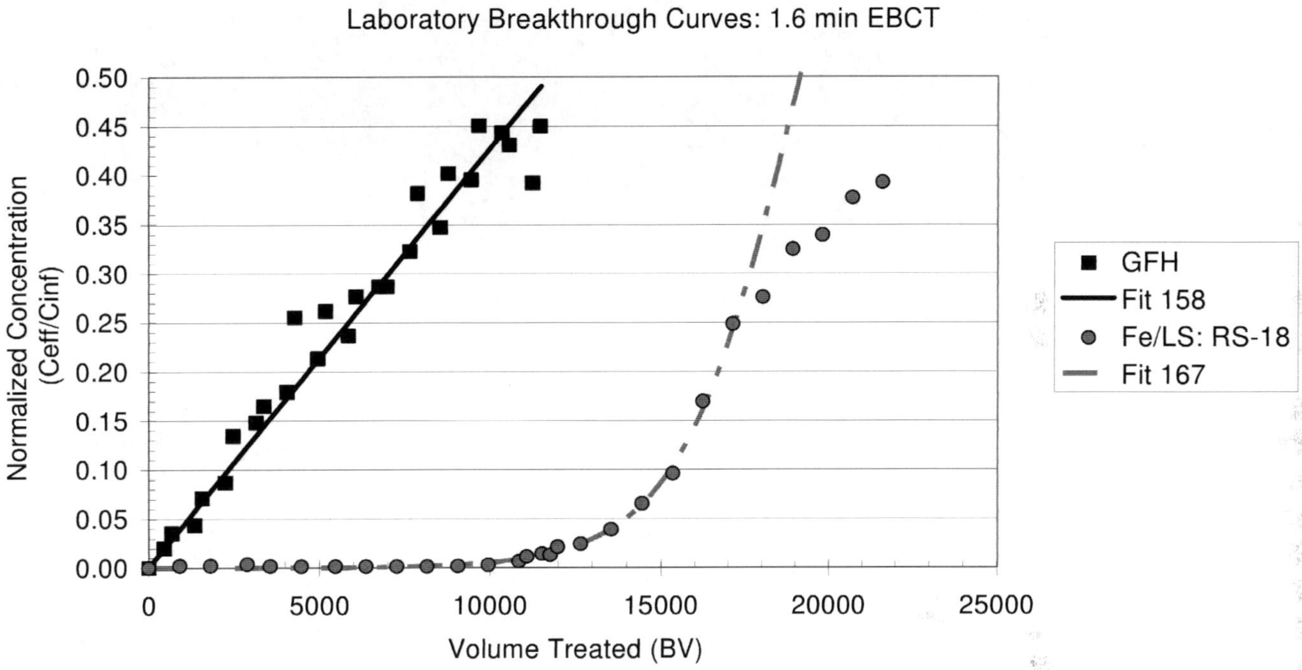

Figure 3. Laboratory Breakthrough Curves. Influent 100 µg/L As(V). EBCT = 1.6 min.

Field Trial Results

The top of Figure 4 shows a photograph of the raw and treated water from the field trial in India. There is clearly a significant improvement of the aesthetic appearance of the water after treatment. Table 4 provides the results from the outside analytical laboratory. After 300 L were treated, the effluent As concentration was non detectable at <1 µg/L, representing >99.3% removal. The total As (both As(III) and As(V)) removal capacity is about 1.5 mg/g media.

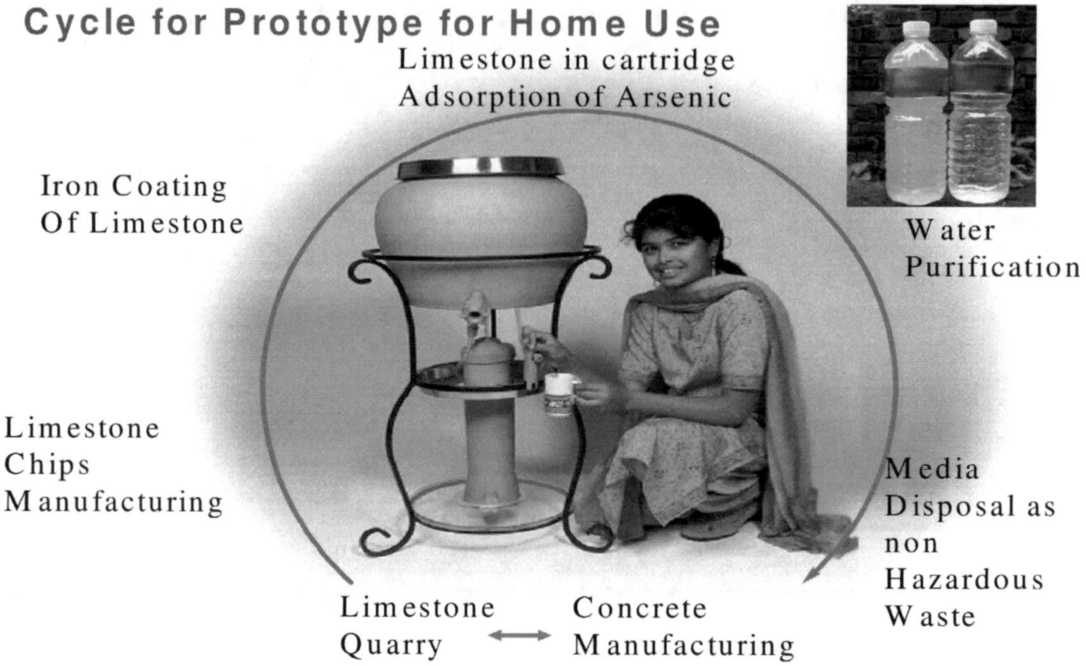

Figure 4: Sustainable Process using Iron Coated Limestone.

Table 4. Field Trial – Water Quality

Test Parameters	Unit	Influent	Effluent (at 300 L)	Removal (%)
Calcium as Ca	mg/L	122	109	10
Magnesium as Mg	mg/L	95	66	30
Iron as Fe	μg/L	7157	1008	86
Arsenic as As	μg/L	136.7	<1.0	>99.3
Manganese as Mn	μg/L	850	160	81
Zinc as Zn	μg/L	320	<70	>80

Not detected: Al, Sb, Cd, Cr, Co, Pb, Ni, Se, Sn and Hg.
Detection Limits: As-1.0 μg/L; Al-500 μg/L; Sb-10 μg/L; Cd-5.0 μg/L; Cr-10 μg/L; Co-500 μg/L; Pb-50 μg/L; Mn-100 μg/L; Ni-90 μg/L; Se-10 μg/L; Sn-10 μg/L; Zn-70 μg/L; Hg-2.5 μg/L.

Theoretical Modeling

The performance of the media was also modeled using the computer program PHREEQC, freely available from the US Geological Survey. This program was designed to model aqueous speciation reactions as well as redox, ion exchange, surface complexation, precipitation, and volatilization reactions for geochemical systems. However, it can also be used to model various ion exchange and adsorption processes in columns.

The modeling served several purposes. It demonstrated that the starting solutions were oversaturated with aqueous CO_2 species relative to atmospheric CO_2 levels, that the solution could spontaneously emit CO_2 to obtain equilibrium, substantially changing the solution pH. The model demonstrated the sensitivity of As adsorption by Fe surfaces to the local pH. The model also demonstrated quantitatively the impact of other species such as SiO_2 and PO_4^{3-} on

As adsorption. Overall, the model was extremely useful in understanding the performance of the new media.

Conclusions

The iron coated limestone has unique properties and has a high capacity for arsenic removal under laboratory and field trial conditions. The unique combination permits selective removal of arsenic through a complexation mechanism and other metals through pH precipitation and Ca complex formation. It is possible to recycle the used limestone into construction materials passing the TCLP method (As < 0.05 mg/L). This leads to an overall sustainable process.

References

1. World Health Organization, http://www.who.int/mediacentre/factsheets/fs210/en/
2. National Academy of Engineering, http://www.nae.edu/nae/grainger.nsf?OpenDatabase
3. Bothe, Jr. J.V., and Brown, P.W., 1999, “Arsenic immobilization by calcium arsenate formation”: Environmental Science and Technology, v. 33, no. 21, p. 3808-3811.

227d Rapid Sensing of Toxic Metals In Water Using a Hybrid Inorganic Material

Prasun K. Chatterjee and Arup K. Sengupta. Civil & Environmental, Lehigh University, 13 E. Packer Avenue, Bethlehem, PA 18015

Rapid and operationally simple sensing of trace concentration of dissolved toxic metals, such as zinc, copper, lead etc in water is desirable for water quality control and mitigation of human health hazard. The present study reveals that a newly synthesized hybrid inorganic material (HIM) is capable of detecting toxic metals, such as zinc, copper, lead and others in water simply through pH changes. HIM is essentially a composite granular material synthesized through rapid fusion of a mixture of amorphous hydrated ferric oxide (HFO) and akermanite or calcium magnesium silicate ($Ca_2MgSi_2O_7$). When a water sample is rapidly passed through an HIM mini-column, effluent pH at the exit always remains alkaline (around 9.0) due to slow hydrolysis of akermanite and steady release of hydroxyl (OH-) ions and at this alkaline pH, HFO offers deprotonated sorption sites for selectively binding toxic metal cations. Commonly encountered electrolytes such as sodium, calcium, chloride show no impact on exit pH. However, presence of trace concentration of toxic metal in water shows a significant pH drop after a specific number of bed volumes. The drop in pH is characteristic of the concentration and the type of toxic metal present. Moreover, the change in the negative slope of pH curve i.e., (-dpH/dBV) provides a sharp, noticeable peak for each toxic metal where BV is bed volumes of solution fed. From a mechanistic viewpoint, high affinity of HFO sorption sites toward toxic metal cations, ability of akermanite to maintain near-constant alkaline pH for a prolonged period through slow hydrolysis and labile metal-hydroxy complex formation provide a synergy that allows semi-quantitative detection of toxic metals at concentrations well below 1 mg/L. According to the information in the open literature, rapid detection of toxic metals using pH as a surrogate parameter has not been reported previously.

227e Magnetic Edta – Attaching Metal Chelators to Nanomagnets for Water Purification and Heavy Metal Recovery

Fabian M. Koehler, Michael Rossier, Evagelos K. Athanassiou, Robert N. Grass, and Wendelin J. Stark. Chemistry and Applied Biosciences, ETH Zurich, Wolfgang-Pauli-Str. 10, ETH Hönggerberg, Zurich, 8093, Switzerland

Magnetic particles are routinely applied in biochemistry separations. Unfortunately their application is often limited due to high costs, poor stability and low binding capacities. Recently we prepared graphene-like coated metallic nanoparticles by reducing flame spray synthesis [1,2]. The coating on the one hand protects the metal nanoparticles from oxidation below 190°C and dissolution in concentrated acids. On the other hand it allows the introduction of covalently bound functional groups via diazonium chemistry [3] or physical adsorption, offering the freedom of organic chemistry on the particle surface.

The functionalized magnetic particles represent a promising modular platform for removing a wide range of contaminants. Attaching EDTA-like chelating agents to the surface allows the removal of heavy metal ions from aqueous solutions. Furthermore, attaching a thiourea-like chelating agent allows the efficient removal of precious metal ions in highly acidic solutions. The low costs and particle stability favor this preparation method and material for large-scale separation application of metal ions in ultra low concentrations.

Figure 1. Magnetic separation test in 10 l water with 1 g carbon-coated cobalt particles. Left figure, start of magnetic separation as a permanent magnet was placed in the lower left corner. Middle figure was taken after 1 minute. After 2 minutes the stirrer was stopped and the right figure shows the end of the separation. More than 97% of the particles were removed.

[1] R. N. Grass, W. J. Stark, J. Mater. Chem. 2006, 16, 1825.

[2] R. N. Grass, M. Dietiker, R. Spolenak, W. J. Stark, Nanotechnology 2007, 18, 035703.

[3] R. N. Grass, E. K. Athanassiou, W. J. Stark, Angewandte Chemie 2007, 26, 4909

Physicochemical Denitrification Process for Drinking Water Resources at Ambient Conditions

Masamichi Tsuji[1], **Mitsuo Kawamura**[2] and **Harue Tsuji**[1],
(1)Aquea Design, Inc., Fujisawa, Kanagawa, Japan,
(2)Tatsumi Industries, Ltd., Kawaguchi, Saitama, Japan

November 16-21, 2008
The AIChE 2008 Annual Meeting
Philadelphia, PA
The American Institute of Chemical Engineers

Nitrogen oxide decomposition to N_2 is economically rational

This is absolutely different from CO_2 decomposition which was proposed in the past.

Benefits of nitrogen oxides decomposition technology:

- N_2O decomposition contributes to mitigation of global warming.
- NO decomposition contributes to removal of air pollutant.
- NO_3^-, NO_2^- decomposition contributes to groundwater purification.

Conventional nitrogen oxides decomposition processes are not sustainable. Why?

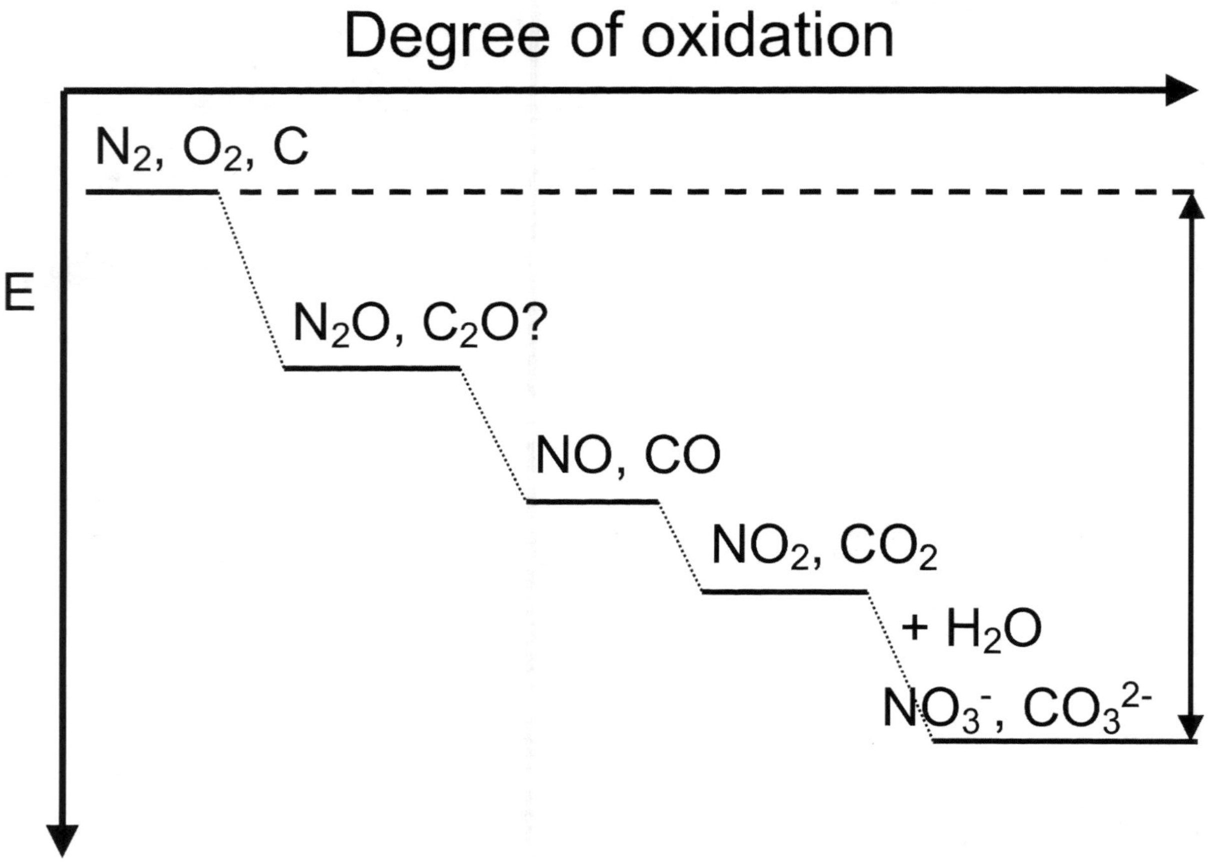

Conventionally this energy was supplied through H_2, noble-metal catalyst or others, resulting in energy-Intensive processes.

Catalytic processes proposed for nitrogen oxide decomposition are “three-body collision”.

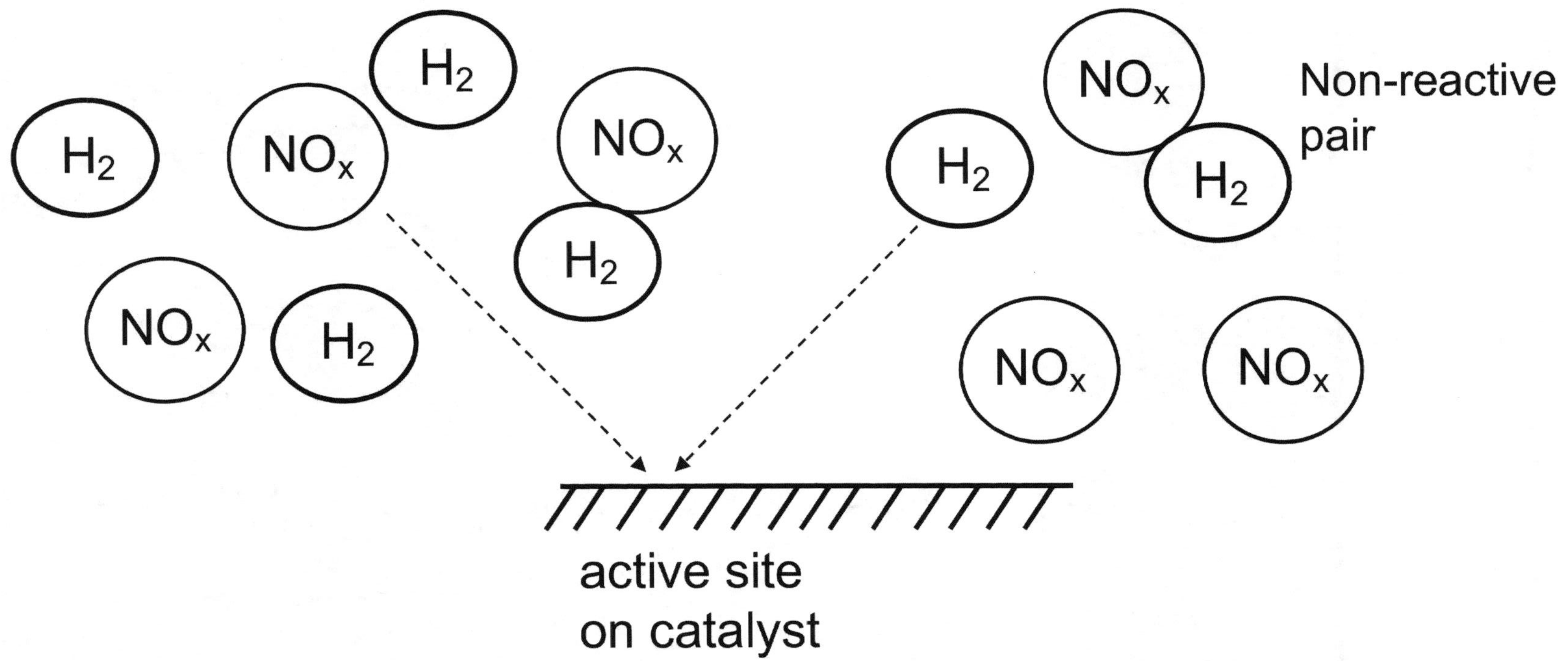

- These are an energy-intensive process requiring lots of high quality energy, e.g., H_2, NH_3 or hydrocarbon.
- Head-on collision of NO_x and H_2 is not sufficient to react at ambient conditions.
- Biological process proceeds at more mild conditions, but too sluggish.

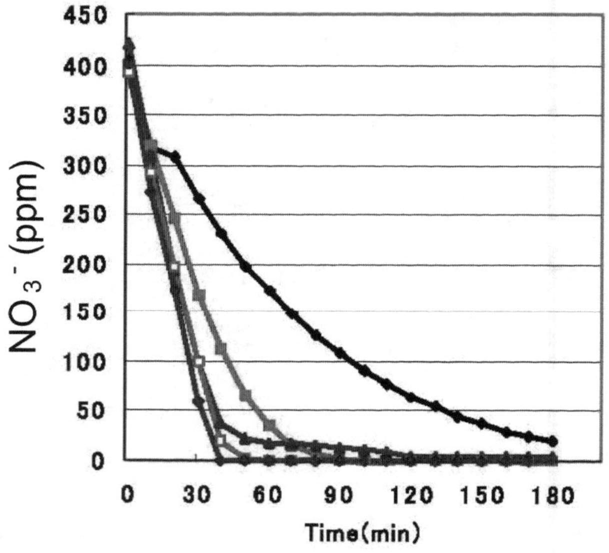

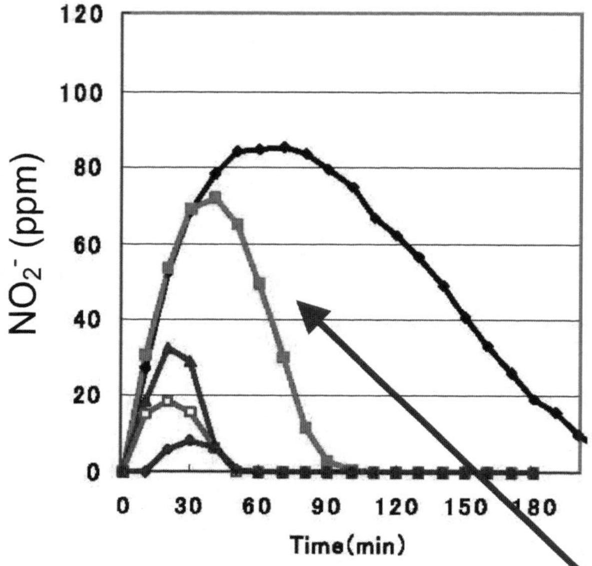

RITE PJ:Catalytic decomposition of NO_3^- in flow of H_2

Results:

Large amount of byproducts : NO_2^- and NH_3

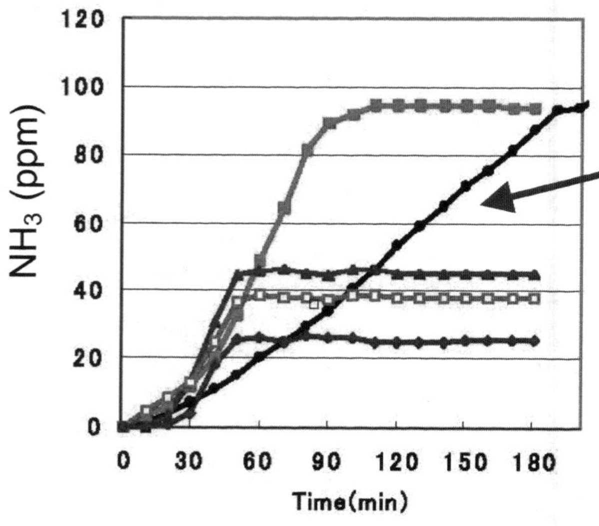

図1-3 NH3量

●0. 05g■0. 15g▲0. 5g□1. 0g★1. 5

Source: RITE report, Japan, March 2005.

Similar unsustainable results have been reported by Hokkaido University and others.

Dr. T. Okuhara (Hokkaido Univ)

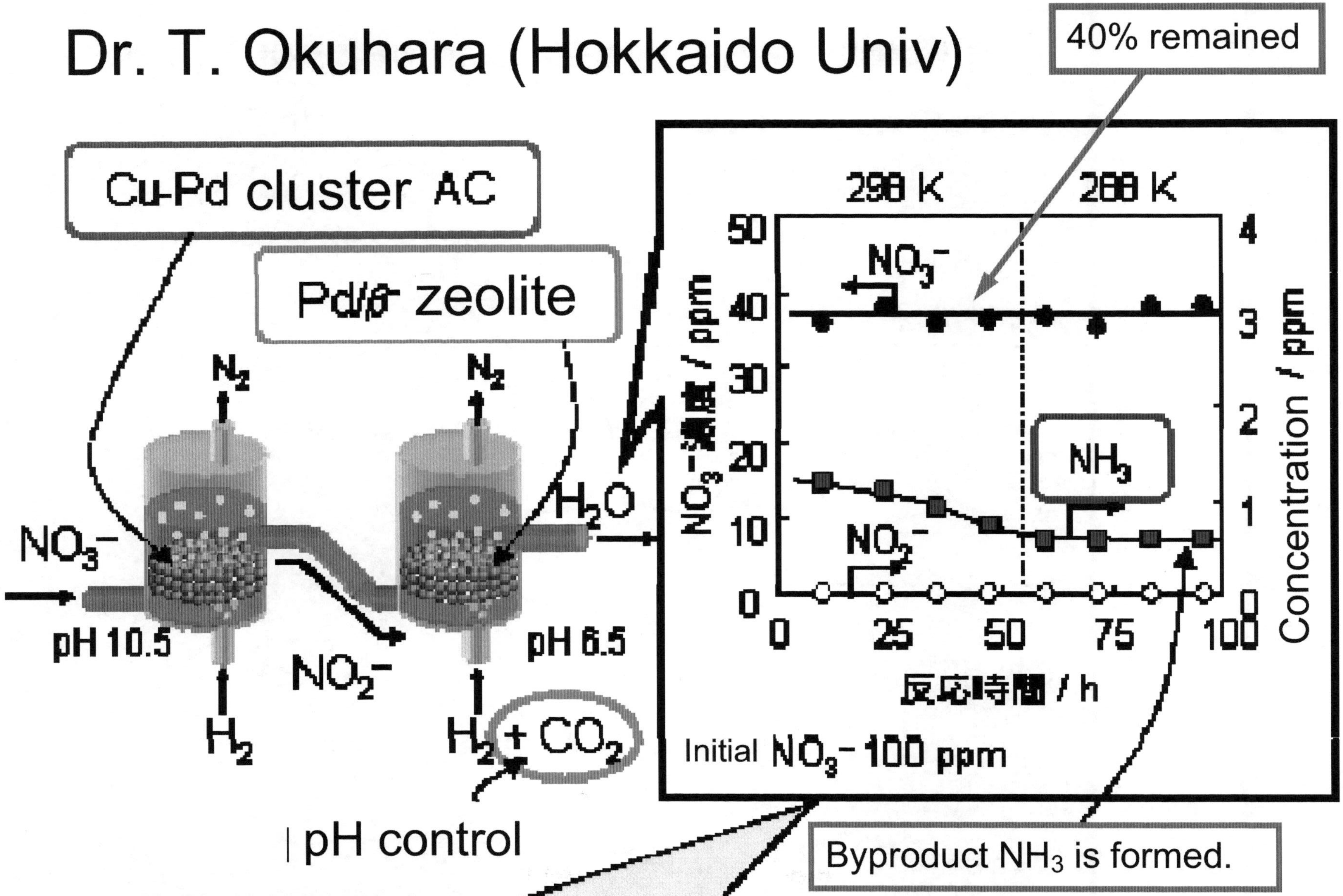

Source: Chem. Lett. 34 (2005) 1510.

Our greener & sustainable idea: spillover-based NO x decomposition

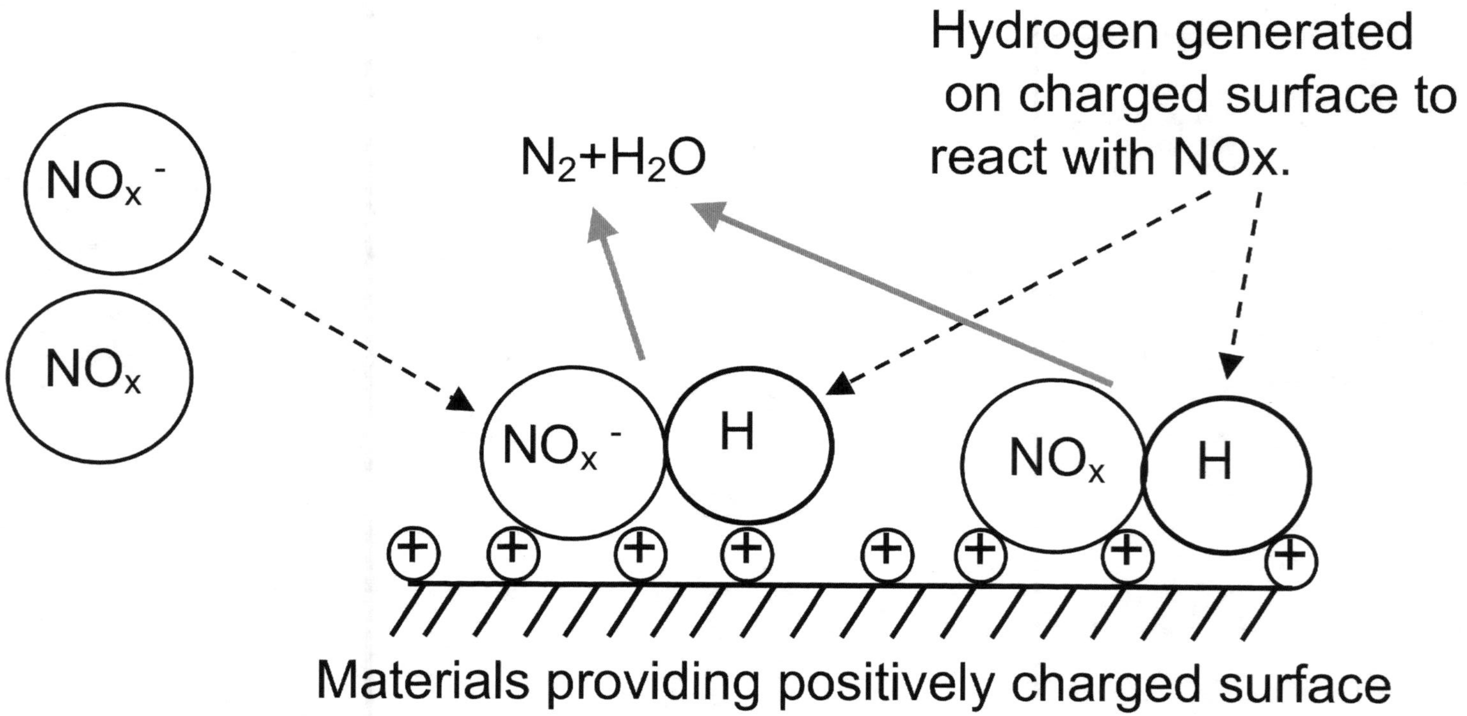

- NO_x decomposition is possible at ambient temperature and pressure.
- Energy input to dissociate NO_x is minimal.

Requisite materials for this greener & sustainable process

Properties of materials:

1. charge-transferable surface
2. serve sufficiently reactive sites even in aqueous conditions.
3. chemically insoluble
4. economical
5. available worldwide
6. non-toxic (for food-level safety, heavy metal free)

Conventional adsorption process

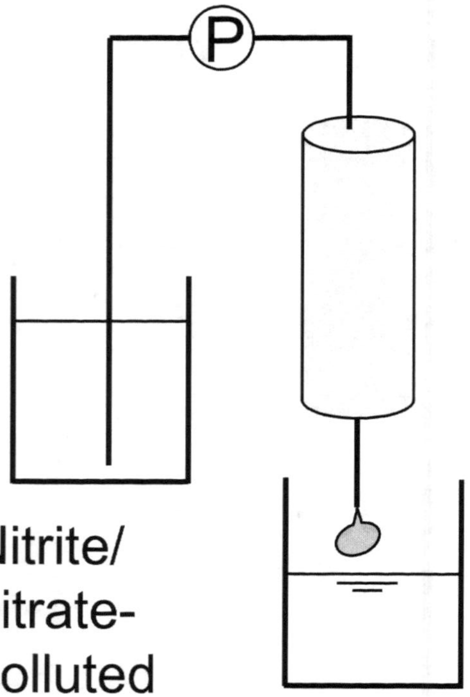

Nitrite/
nitrate-
polluted
water

Cost-effective NOx decomposition process has not been known using conventional adsorption process and well-known materials such as activated carbon and ion exchange resin.

Effluent is not free from NO_2^- or NO_3^-.

Our simple device to generate metabolites-free water with minimal energy at ambient conditions: conceptual drawing

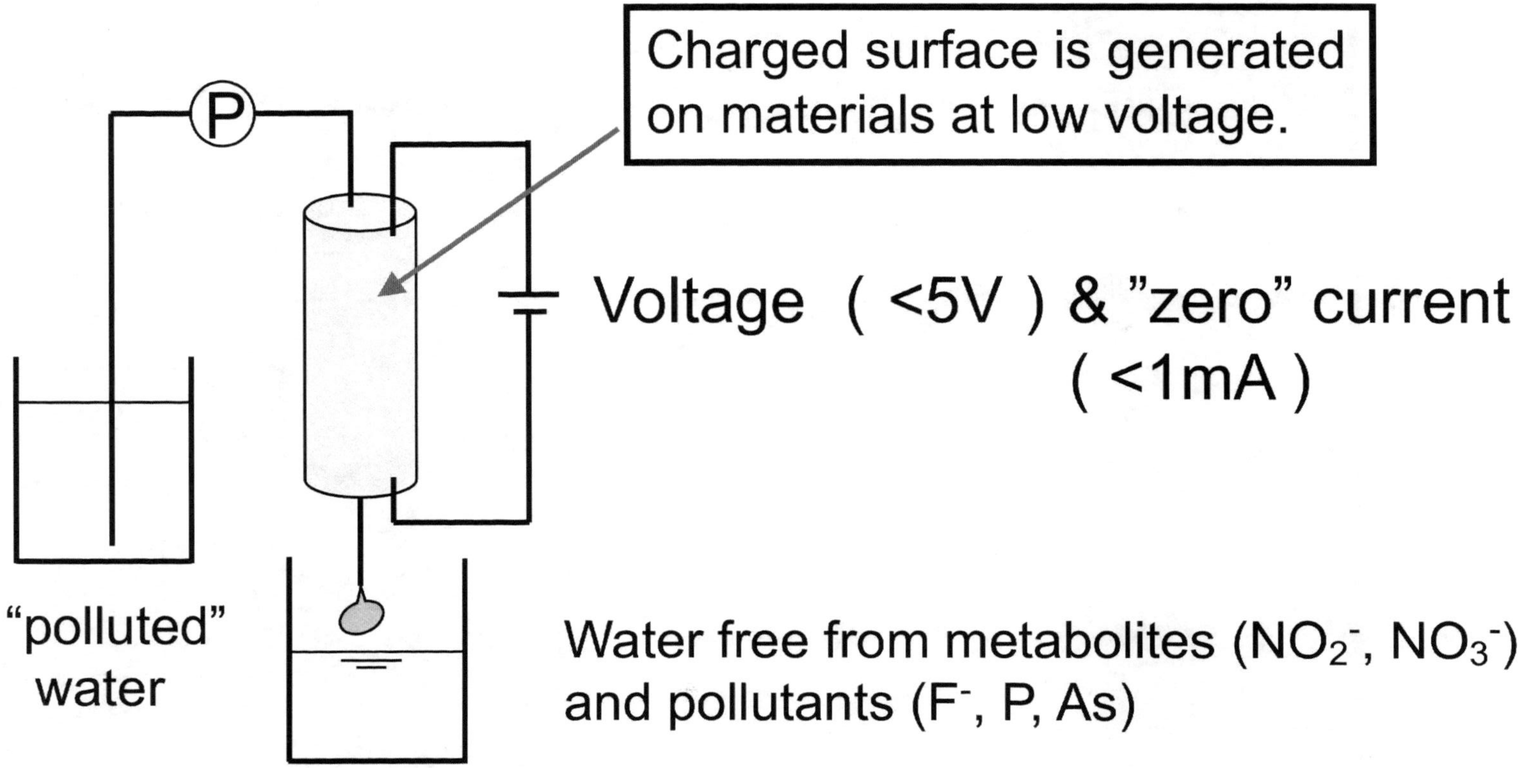

Notes: No use of organics, bacteria, NH_3, or H_2 as reductant of NO_x^-.

Postulated Mechanism for Decomposition of N and S Oxides

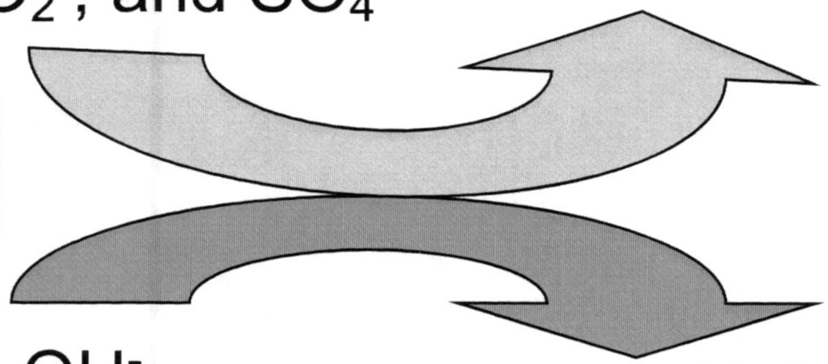

Process at anode : dissociation of water and charge transfer

$H_2O \Rightarrow H + OH \Rightarrow H^+ + OH^-$

- H + NO_3^-, NO_2^- or $SO_4^{2-} \Rightarrow N_2$ or S + H_2O + e^- (decomposition of oxoanions)
- $OH^- + H^+ \Rightarrow H_2O$ (neutralization of acidic water)

Treatment of goldfish bath water

Ions	Feed (ppm)	Feed 1.1L (ppm)	Feed 1.76L (ppm)
PO_4^{3-}	2.95	<0.01	2.39
Cl^-	23.5	97.0	48.5
NO_3^-	143	0.66	0.67
SO_4^{2-}	68.5	5.44	111
pH	3.84	6.90	6.77

Volume of reactor = 0.1L

Total volume of feed = 1.76L

Feeding time = 4 h

SV= 1.76L/4h/0.1L = 4.4

Materials balance of water treatment

Constituent	uptaken (mequiv.)	released (mequiv.)
SO_4^{2-}	1.11	-
HPO_4^{2-}	0.12	-
NO_3^-	4.04	-
Cl^-	-	2.71
Heavy metals	< 0.05	< 0.01
Total	5.27	2.71

Totally different ⇒ not simple ion exchange

(Other anions (CO_3^{2-}, OH^-, H^+) may be involved in this treatment process. Open for study)

FTIR spectrum of material used at anode

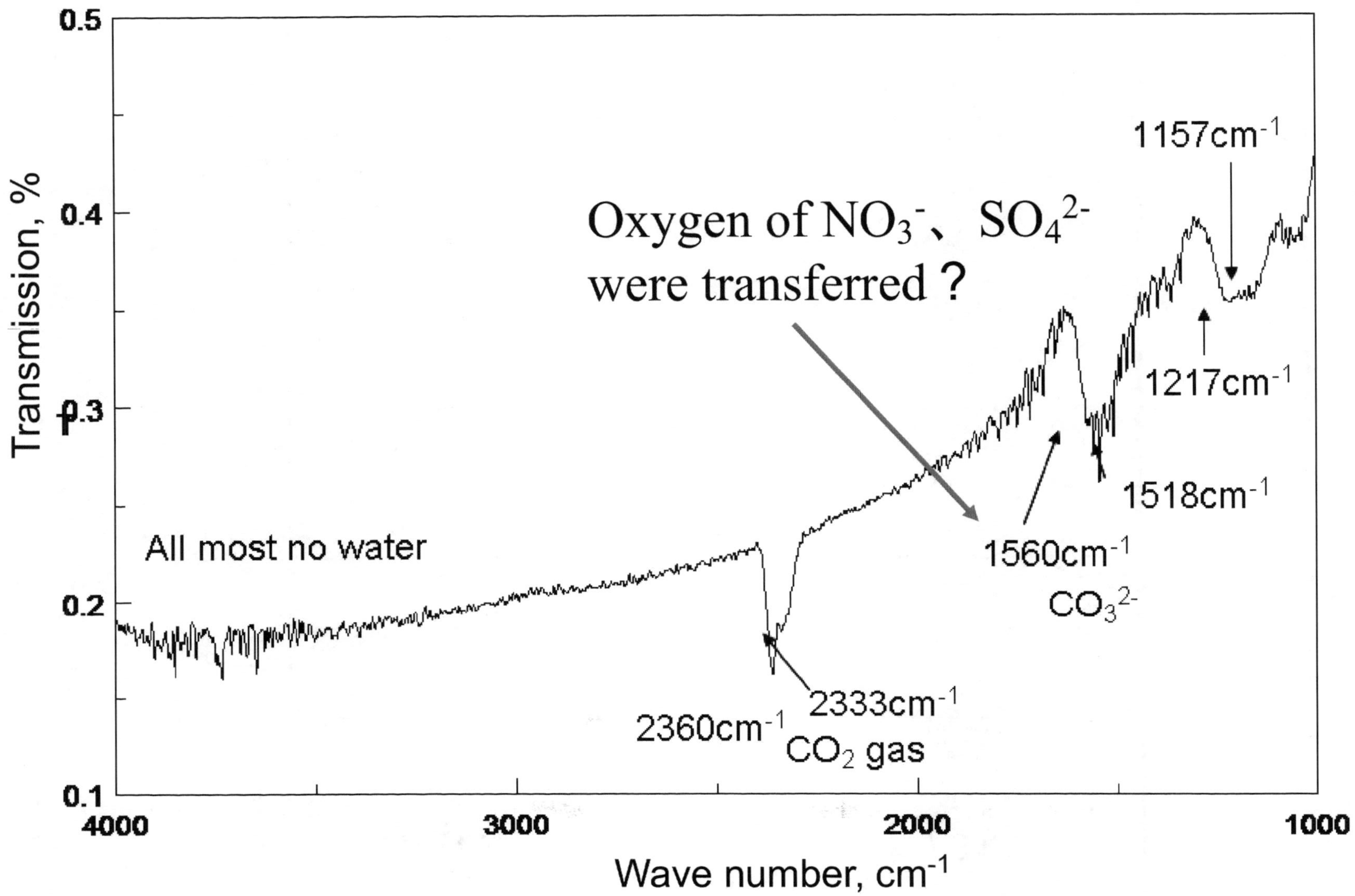

Speciation of sulfur in activated carbon by wave-dispersed PIXE

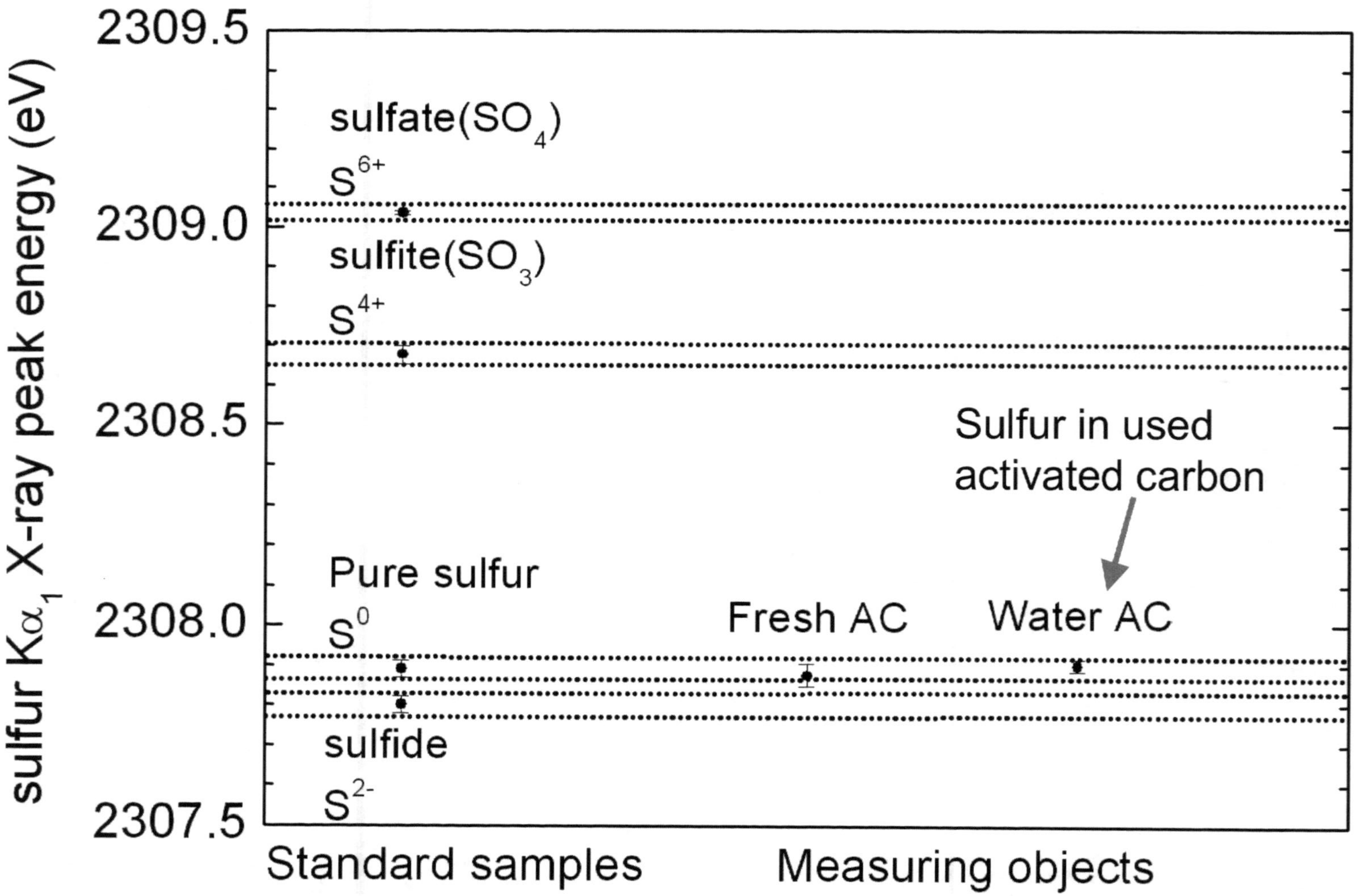

<u>Sulfate was reduced to elemental sulfur.</u>

Conclusion (a)
Operational features

1. Contact time of water with material is very short (<10min).
2. Decomposition of nitrate/nitrite and sulfate is feasible without side reactions to form NH_3.
3. It works at room temperature and at normal pressure.
4. Driving energy is minimal.

Conclusion (b)

Requisite materials in this process

This innovative process does not require:

1. bacteria
2. organic fertilizer or chemicals
3. hydrogen gas

or

4. expensive noble metal-loaded catalysts

Conclusion (c)
Quality of treated water

1. Treatment lowers concentration of NO_3^- from 140ppm to 0.7ppm.
2. Mineral composition of treated water remains the same as before treatment.
3. NH_3 will be never admixed into treated water.
4. Device possesses large neutralization capacity. Acidic groundwater of pH 4 can be neutralized to 7.
5. All the treated water can be utilized. Only 30-40% for conventional reverse osmosis separation method

307a Destruction of Trichloroethylene with Zerovalent Iron/ Persulfate

Anita Padmanabhan and John Bergendahl. Department of Civil Engineering, Worcester Polytechnic Institute, Kaven Hall, 100 Institute Road, Worcester, MA 01609

Research was undertaken to find the potential for zero valent iron (Fe^0)/persulfate treatment of trichloroethylene (TCE) contaminated water. The Fe^0 activated the persulfate to produce sulfate free radicals, a powerful oxidant used for destroying organic contaminants in water. Batch experiments showed that Fe^0/persulfate removed TCE more effectively than persulfate alone, and was more effective than ferrous iron activated persulfate. The results indicated that destruction occurred over a wide range of pH values. The production and destruction of intermediates cis 1,2 dichloroethylene and vinyl chloride was observed. The effect of persulfate dose on this reaction was investigated. Results demonstrated that a molar ratio of 10/1/1 (persulfate/Fe^0/TCE) yielded over 95 percent TCE destruction. Increasing the persulfate dose resulted in greater TCE destruction as well as destruction of the intermediates. Kinetic experiments at a molar ratio of 10/1/1 (persulfate/Fe^0/TCE) showed that approximately 90 percent of the TCE was destroyed in less than 15 minutes.

307b Relationship Between Micellar and Hemi-Micellar Processes and the Bioavailability of Surfactant-Solubilized Hydrophobic Organic Compounds

Derick G. Brown, Department of Civil & Environmental Engineering, Lehigh University, 13 East Packer Avenue, Bethlehem, PA 18015

Surfactants are able to enhance the apparent aqueous concentration of hydrophobic organic compounds (HOC's) through partitioning of the HOC's within surfactant micelles. Many studies have been performed that examined if this effect can result in an enhancement in the biodegradation rate of HOC's. The results have been mixed and the conditions under which enhanced biodegradation may occur remain unknown. One key study demonstrated that a fraction (f) of micellar-phase HOC is directly bioavailable to bacterial cells and a theoretical model for f was provided that gave a good match to a subset of the experimental results, but could not accurately represent all of the surfactants tested. Here, the results of a recent study on surfactant sorption onto the bacterial cell surface are used to describe hemi-micelle formation on the cell surface, and this process is incorporated into the pathway describing micellar HOC bioavailability. The revised model is validated against HOC bioavailability data for five different C12Ey surfactants and it is shown that this model is able to replicate the complete dataset, including that of C12E23 which eluded the original model. The results indicate that surfactant sorption and hemi-micelle formation are important processes governing surfactant-enhanced bioavailability, and these processes are placed into context with other important considerations for surfactant-enhanced biodegradation of HOC's.

307c Metal Oxide Nanoparticles Coated on a Glass Fiber Substrate to Remove Viruses from Water

Xuan Li[1], Leonardo A. Gutierrez[2], Thanh H. Nguyen[2], and James Economy[3]. (1) Department of Materials Science & Engineering, University of Illinois at Urbana-Champaign, 1304 W Green Street, Urbana, IL 61801, (2) Department of Civil and Environmental Engineering, University of Illinois at Urbana-Champaign, 205 N Mathews Avenue. Room 4146, Urbana, IL 61801, (3) Department of Materials Science and Engineering, University of Illinois, 1304 W. Green St, 203 Met. & Min. Bldg. MC-246, Urbana, IL 61801

Microbial contamination of drinking water is a serious problem of global significance. About 51% of the waterborne disease outbreaks in the United States can be attributed to contaminated ground water.

Development of nanoparticles as viricidal materials is of technological and fundamental scientific importance. Nanoparticles with high surface areas and ultra small particle sizes have dramatically enhanced efficiency and capacity of virus inactivation, which cannot be achieved by their bulk counterparts.

A series of metal oxide nanoparticles coated on glass fiber substrates was developed in our research group for evaluation of their viricidal activity.We also carried out XRD, TEM, surface area measurements, and Zeta potential. MS2 virus inactivation experiments showed that these nanoparticle coated fibers were extremely powerful viricidal materials

COMPARISON OF KINETICS FOR REMOVAL OF BTEX COMPOUNDS AND CHLORINATED SOLVENTS USING ADVANCED OXIDATION PROCESSES

ROBERT W. PETERS, Ph.D., P.E.
Professor of Environmental Engineering
Department of Civil, Construction, and Environmental Engineering
University of Alabama at Birmingham
1075 13th Street South
Birmingham, Alabama

M.P. SHARMA, Ph.D.
Professor of Chemical Engineering
Department of Chemical and Petroleum Engineering
University of Wyoming
Laramie, Wyoming

YUSUF GBADEBO ADEWUYI, Ph.D.
Professor of Chemical Engineering
Department of Mechanical and Chemical Engineering
North Carolina Agricultural and Technical State University
1601 East Market Street, 326 McNair Hall
Greensboro, North Carolina

Abstract

This paper addresses the process interactions from combining integrated processes (such as advanced oxidation processes (AOPs), biological operations, air stripping, etc.). AOPs considered include: Fenton's reagent, ultraviolet light, titanium dioxide, ozone (O_3), hydrogen peroxide (H_2O_2), sonication/acoustic cavitation, among others. A critical review of the technical literature has been performed, and the data has been analyzed in terms of the processes being additive, synergistic, or antagonistic. Predictions based on the individual unit operations are made and compared against the behavior of the combined unit operations. The data reported in this paper focus primarily on treatment of petroleum hydrocarbons (e.g., benzene, toluene, ethylbenzene, and xylene) and chlorinated solvents. The kinetic data included in this study are obtained both from the technical literature and from research investigations of authors. The results of this study are discussed in this paper. Our first order kinetic rate constants are comparable with other researchers, although our values seem to be slightly higher than the previously reported values. Our air stripping values are significantly higher than the volatilization rate constants previously reported; this is due to the earlier study conducting volatilization studies under quiescent conditions.

Introduction

Description of Advanced Oxidation Processes

Acoustic cavitation (commonly termed sonication) involves the application of sound waves being transmitted through a liquid as a wave of alternating cavitation cycles. Compression cycles exert a positive pressure on the liquid, pushing molecules together, while expansion cycles exert a negative pressure, pulling molecules away from each other. The chemical effects of sonication are a result of acoustic cavitation. During rarefaction, molecules are torn apart forming tiny microbubbles that grow to a critical size during the alternating cavitation cycles, and then implode releasing a large amount of energy. Temperatures on the order of 5,000°K and pressures up to 500 to 1,000 atmospheres have been observed in microbubble implosions, while the bulk solution stays near ambient. The collapsing bubble interface results in the formation of hydroxyl and hydrogen radicals. These radicals destroy chlorinated organic and petroleum hydrocarbon compounds very effectively.

The performance can be further enhanced through the addition of advanced oxidants (e.g., ozone (O_3), hydrogen peroxide (H_2O_2), etc.), and incorporation of recent advancements in the acoustic cavitation field [1]. The destruction of organic pollutants can occur via several mechanisms. The organic pollutant inside the cavity and in the interfacial region can undergo pyrolysis reactions (or combustion reactions if oxygen is present) during the implosion. Free radicals (e.g., •OH, •H) formed due to thermolysis of water molecules may react with the organic in the interfacial region or in the solution near the interface. Three primary pathways have been identified for compound degradation, including: (1) hydroxyl radical oxidation, (2) direct pyrolytic degradation, and (3) supercritical water reactions. In aqueous solution, water vapor present in the microbubble is homolytically split during bubble collapse to yield •H and •OH radicals, while chemical substrates present either within or near the gas-liquid interface of the collapsing microbubble are subject to direct attack by •OH. Volatile compounds such as benzene, toluene, ethyl benzene, xylene, and MTBE readily partition into the vapor of the growing cavitation microbubbles and then undergo direct pyrolysis during transient collapse.

Vapor stripping (air sparging) operations have as their goal to transfer the volatile contaminants from the liquid phase (i.e., groundwater) to the vapor phase (i.e., air). The combination of sonication and aeration/ozonation/hydrogen peroxide results in enhanced organic destruction and better liquid/vapor contact (i.e., better mixing) to facilitate the transfer of the degraded or partially degraded products into the gas phase. The addition of ozone (made through electric discharge of oxygen) into the liquid phase coupled with application of sonication considerably enhances the formation of hydroxyl radicals resulting in enhanced destruction of chlorinated organic compounds in solution. Hydroxyl radicals (•OH) are highly reactive non-specific reactants capable of oxidizing a wide variety of contaminants to carbon dioxide, hydrochloric acid (if chlorinated organics are present), and water. After mass transfer from the liquid phase into the gas phase, the water is pumped to a packer in the dual extraction well, resulting in the water (stripped of the volatile organic compounds (VOCs)) being returned into the subsurface, while the VOC-enriched vapor phase is passed through the second well and is sent to an above-ground vapor treatment system (that could involve capture on activated carbon/regeneration using ultrasonics, or thermal oxidation).

This research paper considered the following advanced oxidation treatment systems:

- Ultraviolet (UV) light;
- Hydrogen peroxide (H_2O_2)/UV light;
- Sonication;
- Sonication/UV light;
- Ozonation;
- Ozonation/H_2O_2; and
- Fenton's Reagent.

Previous Studies with Evidence of Process Synergism

The photooxidation of *p*-xylene and *o*-nitrotoluene were studied by Ghaly [2] in a closed bench-scale unit using hydrogen peroxide or hydrogen peroxide/Fe^{+2} as a photocatalyst. Results of the photo-Fenton process (UV/H_2O_2/Fe^{+2}) were compared with the UV/H_2O_2 process; the results indicated that the photo-Fenton process resulted in a higher efficiency degradation of the selected compounds and reduced energy consumption. The degradation reactions were 1st -order; the value of the reaction rate constant k was ~5x higher for the photo-Fenton process than that in the case of the UV/H_2O_2 system. The quantity of both H_2O_2 and Fe^{+2} must be properly selected the pH properly adjusted to optimize the process. For their particular system studied, the optimum reaction conditions were: 0.03 mole H_2O_2/L using the UV/H_2O_2 system and 0.01 mole H_2O_2/L, 1.0 mmole Fe^{+2}/L at pH ~3 for the photo-Fenton process.

The ozonation and peroxone oxidation of toluene in aqueous solutions was studied by Kuo and Chen [3]. They investigated the kinetics of the aqueous-phase oxidation or toluene by ozone and ozone-hydrogen peroxide mixtures at 25°C. The oxidation kinetics were 1st-order with respect to ozone concentration; the reaction order in toluene varied with pH and the presence or absence of H_2O_2. The peroxone oxidation was half-order with respect to H_2O_2 in distilled water (initial pH of 5.4) and other solutions of higher pH. In acidic solutions with an initial pH $\leq$ 3, the overall kinetics was 2nd-order; the direct oxidation of toluene by ozone molecules was predominant in determining the slow rate of reaction. The reaction became very fast and enhanced with the addition of H_2O_2, if present, in alkaline solutions with an initial pH of 10 or higher. Under those conditions, the reaction was controlled by hydroxyl radical reactions and was independent of the toluene concentration.

RESULTS AND DISCUSSION

Previous Studies Involving Combinations of Advanced Oxidation Studies

Figure 1 presents the results on the removal of atrazine from solution [4] using UV irradiation, ozonation, and combined UV/ozonation, while **Figure 2** presents analogous results for removal of carbofuran [5] using UV irradiation, Fenton's reagent, and combined UV/Fenton's treatment. **Figure 1** indicates that the removal of atrazine from solution is better than being additive, that much better performance is observed than predicted, indicating the process performed synergistically. Similar behavior is observed for the case of carbofuran. Beltran *et al.* [4] showed that the degradation efficiencies of atrazine by advanced oxidation treatment methods were in the following order: UV < O_3 < UV/H_2O_2 < UV/O_3. Similarly, Benitez *et al.* [5] observed that the degradation efficiencies of carbofuran were in the following order: UV < O_3 < Fenton's reagent < UV/H_2O_2 < UV/Fenton. Note that the integrated processes always had the best removal efficiencies of the target compounds.

Figures 3 and 4 present the results for removal of carbon tetrachloride (CCl_4) and trichloroethane (TCA) from solution using combined sonication and air stripping operations [6, 7]; the prediction from the additivity of the 1st-order rate constants is shown by the curve in the figure; the actual observed data is shown as the individual data points. Once again, the combination of the two treatment processes (sonication and air stripping) is seen to operate in a synergistic fashion.

Tables 1 and 2 compare the 1st-order rate constants using sonication, air sparging, and their combination for removal of benzene, toluene, ethyl benzene, and *o*-xylene from solution [1]. Once again, the combination of these treatment processes is considerably better than the individual processes, and operates in a synergistic fashion, as evidenced by the observed 1st-order rate constants being higher than the predicted values.

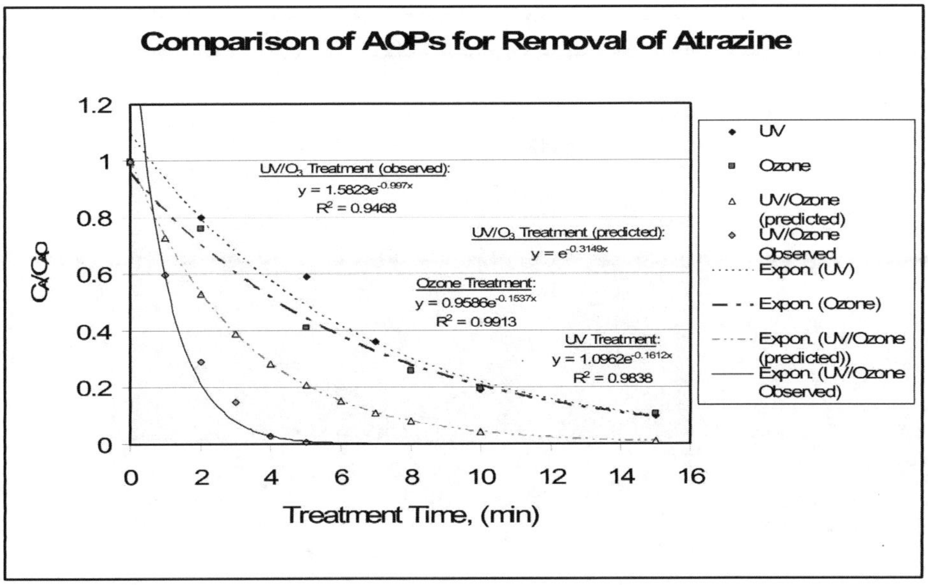

Figure 1. Comparison of AOPs for removal of atrazine from solution, adapted from [2].

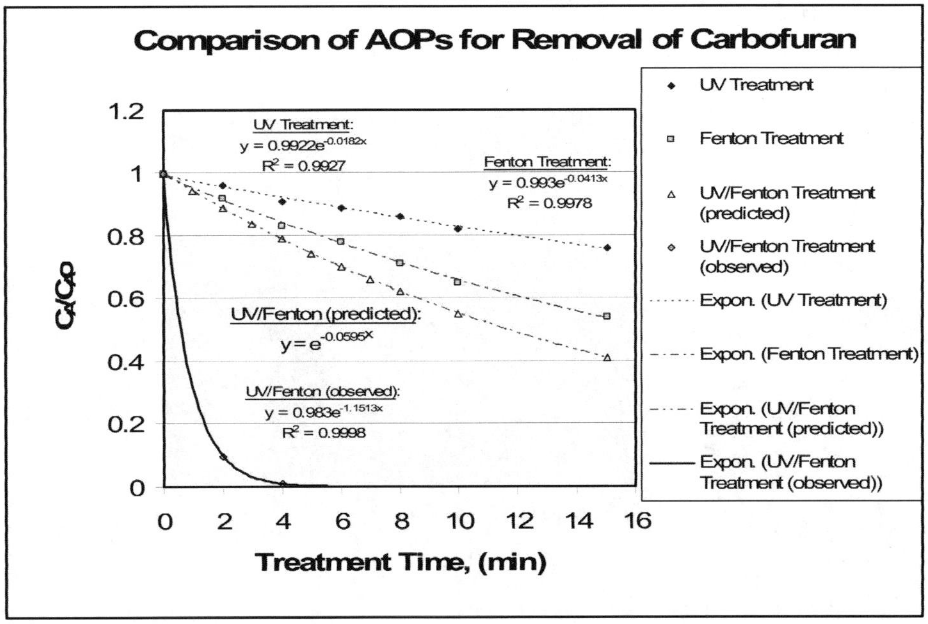

Figure 2. Comparison of AOPs for removal of carbofuran from solution, adapted from [3].

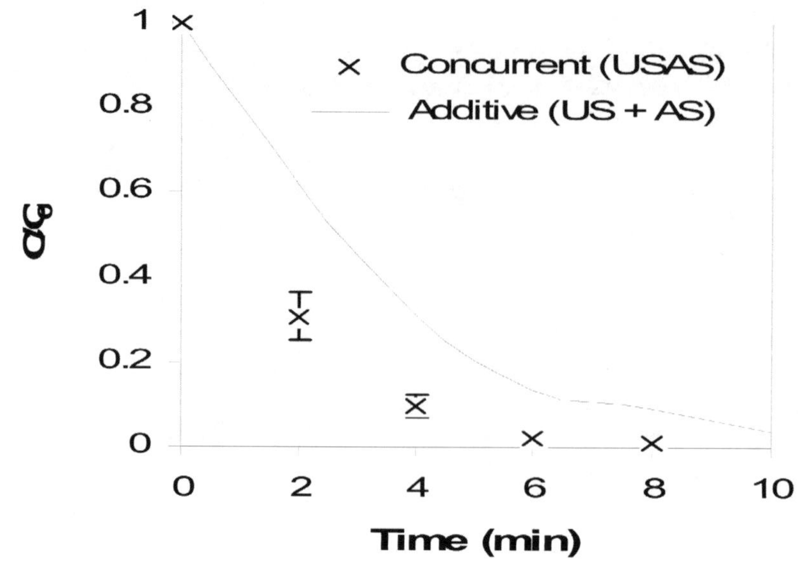

Figure 3. Removal of CCl_4 from artificial groundwater by combined ultrasound and air stripping: Ultrasonic power intensity = 35 W/cm^2 and air flow rate = 500-mL/min.

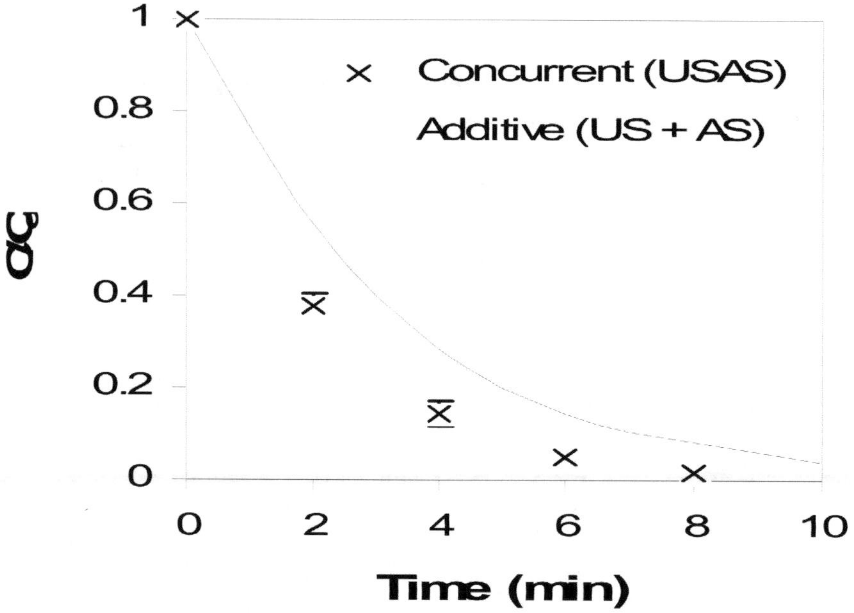

Figure 4. Removal of TCA from artificial groundwater by combined ultrasound and air stripping: Ultrasonic power intensity = 35 W/cm^2 and air flow rate = 500-mL/min.

Table 1. Summary of 1st-order rate constants using various treatment processes for removal of benzene, toluene, ethyl benzene, and *o*-xylene from solution [1].

Contaminant	Treatment Process	1st-Order Rate Constant, (min^{-1})
Benzene	Sonication alone	0.0608
	Air sparging alone	0.1347
	Combined sonication + air sparging (predicted)	0.1955
	Combined sonication + air sparging (observed)	0.2166
Toluene	Sonication alone	0.0635
	Air sparging alone	0.1381
	Combined sonication + air sparging (predicted)	0.2016
	Combined sonication + air sparging (observed)	0.2443
Ethyl benzene	Sonication alone	0.0702
	Air sparging alone	0.1655
	Combined sonication + air sparging (predicted)	0.2357
	Combined sonication + air sparging (observed)	0.3684
o-Xylene	Sonication alone	0.0497
	Air sparging alone	0.1385
	Combined sonication + air sparging (predicted)	0.1882
	Combined sonication + air sparging (observed)	0.2132

Table 2. 1st-Order Kinetic Rate Constants Obtained by Other Researchers.

Contaminant	Treatment Process	1st-Order Rate Constant, (min^{-1})	Reference Citation
Benzene	Sonication with a Near Field Acoustic Processor for applied power density of 0.6–3.6 W/mL	0.0027 – 0.0374	Thoma *et al.*, 1998
Benzene	Sonolysis	0.025	Wu *et al.*, 2004
Benzene	Volatilization @ 75-mm depth	0.000656	Peng *et al.*, 1994
Toluene	Sonication with a Near Field Acoustic Processor for applied power density of 0.6–3.6 W/mL	0.0039 – 0.0276	Thoma *et al.*, 1998
Toluene	Volatilization @ 75-mm depth	0.000692	Peng *et al.*, 1994
1-dodecanol	KOH alcohol polyethoxylation	0.00018 – 0.00348	Santacesaria *et al.*, 1992

Summary and Conclusions

For the research investigations studied, integrated processes generally perform in a synergistic manner, rather than giving the performance of the individual processes simply being additive. For example, a synergistic effect is observed in the photo-Fenton process which was attributed to the generation of hydroxyl radicals (•OH) in several reactions and to the catalytic behavior of the Fenton's reaction in the presence of UV radiation. Thus, for the oxidant concentrations employed in the Benitez study [5], the efficiency of the photo-Fenton process was even higher than that of the conventional ozonation and the UV/O_3 system. As shown in the analyses presented in this paper, the rate constant of the 1st-order combined treatment system was 3.167x and 19.35x greater than predicted rate constant (the sum of the individual unit processes rate constants) for the removal of atrazine and carbofuran, respectively. For the removal of CCl_4 and TCA [6, 7], the observed rate constant was 1.777x and 1.653x times greater than the predicted value (the sum of the 1st-order rate constants for sonication and air stripping), respectively. Similarly, for treatment of benzene, toluene, ethyl benzene, and *o*-xylene, the observed rate constant [1] was 1.11x, 1.21x, 1.56x, and 1.13x times greater, respectively, than the predicted value (the sum of the 1st-order rate constants for sonication and air stripping). Sonication by itself is not a highly effective process, while combined processes (e.g., sonication + air sparging, sonication + H_2O_2, sonication + UV, etc.) are synergistic in nature, providing much greater treatment effectiveness than the efficiencies predicted by the sum of the 1st-order rate constants. The improvement in rate constants resulting from integrated processes can be an order-of-magnitude or more high than those of single processes alone (such as sonication or air sparging).

References

[1] Peters, R.W., Gauthier, J.J., and Mohammad, J., (2006). "Final Report: Use of Sonication/Acoustic Cavitation with Advanced Oxidants to Treat Petroleum Hydrocarbons-Contaminated Surface Waters and Groundwaters", Report submitted to the Alabama Water Resources Research Institute.

[2] Ghaly, M.Y., (2005). "Destruction of the Environmental Organic Pollutant through Activated Hydrogen Peroxide Oxidation", *Afinidad*, *62*(518): 314-320.

[3] Kuo, C.-H., and Chen, S.-M., (1996). "Ozonation and Peroxone Oxidation of Toluene in Aqueous Solutions", *Indus. Eng. Chem. Res.*, *35*(11): 3973-3983.

[4] Beltran, F.J., Garcia-Araya, J.F., and Acedo, B., (1994). "Advanced Oxidation of Atrazine in Water – II. Ozonation Combined with Ultraviolet Radiation," *Water Research*, *28*(10): 2165 – 2174.

[5] Benitez, F.J., Acero, J.L., and Real, F.J., (2002). "Degradation of Carbofuran by Using Ozone, UV Radiation, and Advanced Oxidation Processes", *J. Hazard. Mater.*, *89*(1): 51-65.

[6] Peters, R.W., and Sharma, M.P., (2006). "A Critical Evaluation of Integrated Advanced Oxidation Processes: Assessment of Process Synergism", Paper presented at the 2006 AIChE Meeting, San Francisco, CA, (November 12-17).

[7] Ayyildiz, O., (2003). "Ultrasonic and Air Stripping Removal of CCl_4 and TCA from Water: Investigating Synergistic Effects", Ph.D. Dissertation, Illinois Institute of Technology, Chicago, IL, (July).

[8] Peters, R.W., Sharma, M.P., Adewuyi, Y., "Combinng Advnaced Oxidation Processes: Assessment Of Process Additivity, Synergism, and Antagonism", Paper presented at the 2007 AIChE Spring National Meeting, (March).

Disinfection Byproducts Control with Ion Exchange

Francis Boodoo [a]
DJ Shannahan [b]
Edward Begg [a]
Joe D'Alessandro [a]

[a] The Purolite Company
[b] Sharpe Water

Abstract

Reduction of TOC prior to chlorination is an effective way to control the formation of Disinfection Byproducts (DBP). Use of brine regenerable ion exchange resins for TOC removal is an established and cost effective technology. Minimizing brine discharges to the environment is now very important in arid states as more municipal water treatment plants require the waste water from brine regenerable ion exchange systems be hauled away for disposal. Operating costs for such systems are now 4 to 6 times higher than for comparable systems with onsite brine disposal facilities. This presentation gives an overview of current technologies and explores a multiple reuse strategy for the spent brine from ion exchange systems, thereby reducing waste and reducing operating cost for the process.

Keywords: Disinfection byproducts, TOC, ion exchange, brine reuse

1.0 Introduction

Ion exchange resins occupy a unique and prominent position among technologies for potable and ground water treatment. Their capacity for trace contaminant removal and their ability to be regenerated and reused for hundreds of cycles are main contributors to their comparatively low operating costs versus other technologies. It is not surprising therefore that ion exchange resins are used for a host of purification applications, including the removal of hardness, arsenic, nitrate, uranium, perchlorate and organic matter (TOC) from potable water.

Disinfection of water containing TOC with oxidants such as chlorine and ozone can result in the formation of disinfection products (DBPs) such as trihalomethanes, haloacetic acid and bromate at levels exceeding their current maximum contaminant levels (MCLs) of 80 ppb, 60 ppb, and 10 ppb respectively. The recent promulgation of the Stage 2 DBP rule by the US EPA tightens operator control over pathogens (e.g. bacteria, cryptosporidium and giardia) as well as DBPs by requiring additional monitoring at individual points in the water distribution system. Consequently, organic matter control prior to the point of oxidant dosing has become even more critical.

2.0 Reducing TOC

A variety of methods are currently being used for TOC reduction from drinking water including:

- Coagulation / Filtration
- Granular Activated Carbon (GAC)
- Coagulation / Filtration + GAC
- Microfiltration (MF) + Coagulant Feed
- MIEX pretreatment
- Brine Regenerable Strong Base Anion Exchange Resins with option for brine reuse

A typical treatment scheme using coagulation / filtration followed by polishing with granular activated carbon is shown in Figure 1 below.

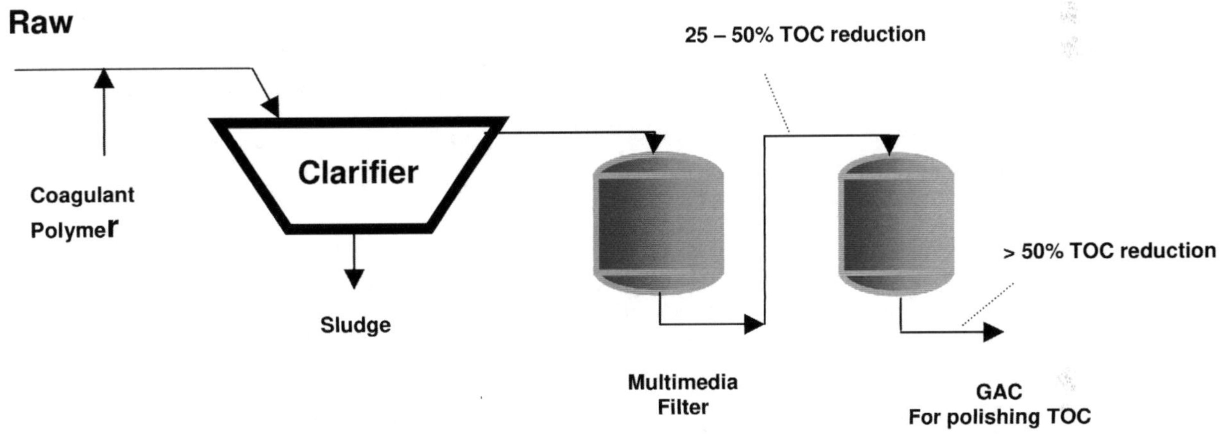

Figure 1. Typical Coagulation / Filtration + GAC Scheme

Clarification typically removes about 25% to 50% of the TOC depending on the characteristics of the TOC, while the addition of GAC as a polisher usually improves on this further.

Microfiltration (MF) coupled with the dosing of a suitable coagulant upstream of the MF unit has become popular in recent years. However, TOC removal claims vary widely, ranging from 10% to90%, with typical values more likely to be in the range of 10% to 30%. Figure 2 shows a typical treatment scheme.

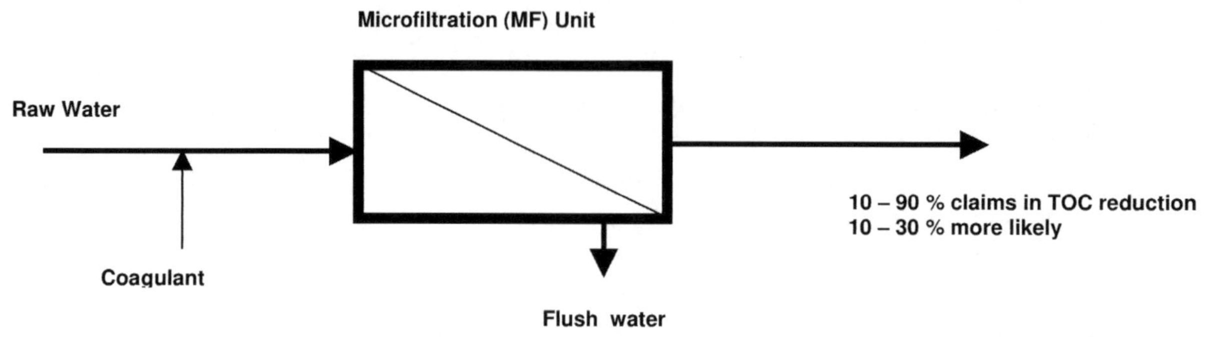

Figure 2. Typical Microfiltration (MF) + Coagulant Treatment Scheme

TOC removal using a variety of anion exchange resins has been used for many years. It is possible to get reliable and predictable results if the ion exchange plants incorporating such resin are properly designed and operated. Such ion exchange plants are relatively easy to operate and require only a simple brine solution to regenerate the resin. The ability to use the same standard fixed bed ion exchange equipment and controls as those used for softening, and the ability to efficiently regenerate these resins makes them very cost-effective options for control of organic matter. Typical operating conditions are shown in Figure 3 below.

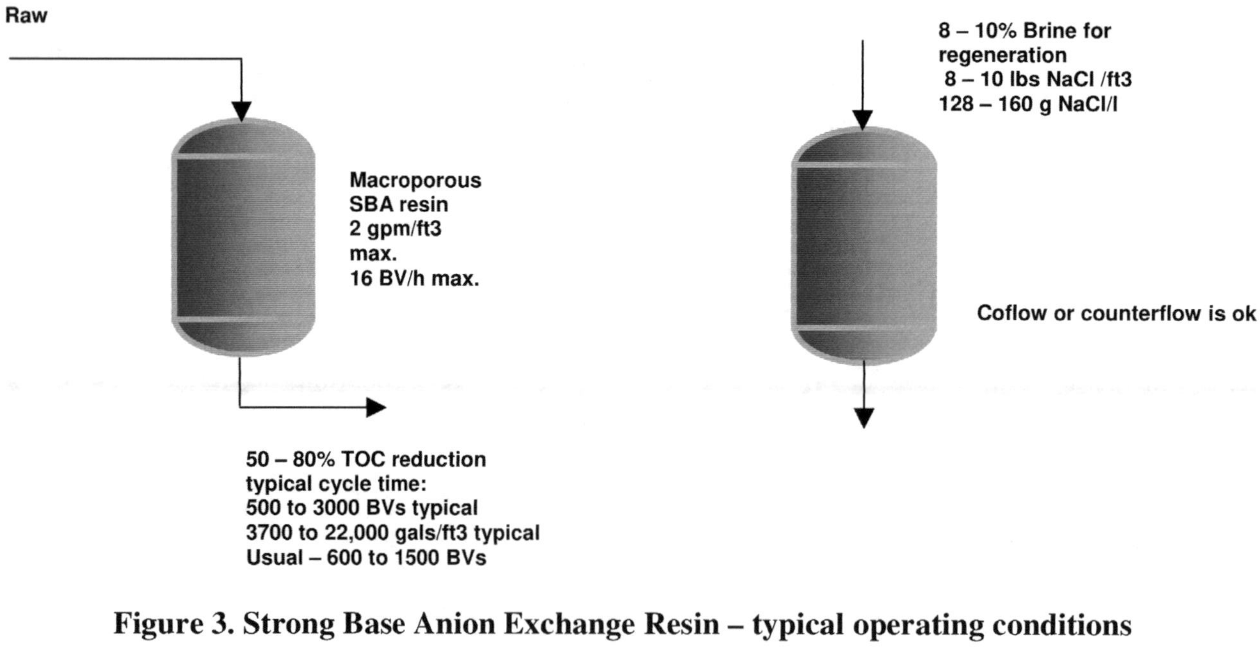

Figure 3. Strong Base Anion Exchange Resin – typical operating conditions

A relatively new ion exchanged based process called MIEX has also gained recognition in recent years in the US. This ion exchange based process uses an iron-infused strong base anion resin for TOC removal in a custom designed treatment process as shown in Figure 4.

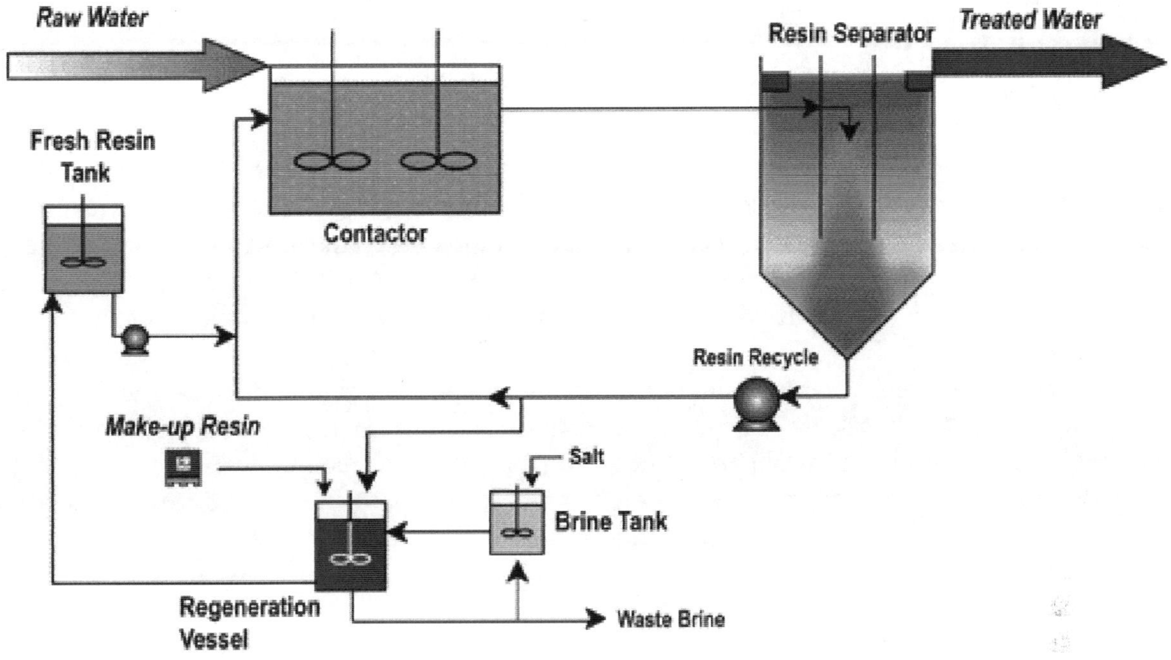

Figure 4. Typical MIEX Treatment Scheme (Courtesy MIEX)

The major difference from the standard ion exchange process is the use of relatively small iron-infused resin beads in a fluidized environment, with typical diameters raging from 150 to 180 microns as compared to the standard process which uses a Gaussian distribution of beads with diameters ranging from 300 to 1200 microns. The water and beads are mixed together in an open contactor tank, allowing quicker adsorption of the organic matter on the smaller beads. The beads are then separated from the treated water, taking advantage of the higher density caused by the iron infusion, then they are routed for regeneration with brine regeneration before being returned to the contactor tank. Fresh resin is continually added to make up for losses in resin volume due to the higher rate of bead attrition arising from the constant movement and impacts.

2.1 Design Considerations for Ion Exchange Systems

The extent of TOC removal by ion exchange depends on the nature of the organic matter in the water to be treated and the chemical characteristics of the resin itself. Both styrenic and acrylic type resins can be used. Naturally occurring organic matter (NOM) is comprised to a large extent of humic, fulvic and tannic acids. The majority of compounds tend to be non-polar and hydrophobic and can be adsorbed by either carbon or hydrophobic type resins[1].

The rest, or the polar fraction of the NOM is characterized as being non-adsorbable on carbon and are thought to be composed of low molecular weight compounds (< 1000) that are "water-loving" or hydrophilic in nature. Since strong base anion resins possess both adsorption and ion exchange properties, these resins can pick up both the polar and non-polar compounds. This is a fundamental difference and advantage compared to activated carbon picks up very little of the polar compounds.

More importantly, from a DBP control standpoint, research has shown that the compounds in raw water that can later form trihalomethanes (THM precursors) are evenly split between polar and non-polar fractions. This means that both polar and non-polar compounds must be removed before oxidation to assure complete control over DBP formation. Ion exchange resins therefore offer a more comprehensive solution for both types of organic matter.

In selecting a regenerable resin, good organic adsorption characteristics must be matched with good efficiency of release during regeneration. Hydrophobic styrenic resins form strong bonds with non-polar organics that are difficult to break during regeneration. They are especially susceptible to fouling by high molecular weight organic compounds. Acrylic resins, on the other hand, can pick up both polar and non-polar organic compounds. The hydrophilic nature of acrylic resins results in easier release of the organics during regeneration and is the preferred choice in many cases.

Good design rules are equally important to get the maximum benefits from the system. Most drinking water systems are designed to use just brine for regeneration, using about 8 to 10 pounds of salt per cubic foot of resin (or 128 to 160 grams NaCl per liter of resin) and typical service flow rates of 1.5 to 2 gpm/ft3 (12 to 16 BV/h). Regeneration in co-flow or counterflow modes are both acceptable. For difficult jobs, adding a small amount of caustic (where permitted) will generally help with the smoother release of the organics from the resin matrix, but the impact on the chemistry of the treated water must first be assessed.

2.2 Field Pilot Results for Brine Reuse:

Recovery and reuse of the spent brine for subsequent regenerations has been well documented in previous studies[3] for TOC removal, with successful reuse of the brine for as many as eight times before disposal. Thus the current fieldwork explores the use of this technique for a specific water chemistry and verifies regeneration efficiency from cycle to cycle in terms of residual TOC on the resin.

In attempting to recover and reuse the spent brine from such systems, it is important to understand the role of the brine in regenerating off TOC from the resin. Since the organic matter is adsorbed onto the resin matrix by both adsorption and ion exchange mechanisms, the brine serves two purposes. In one instance the brine simply displaces the relatively bulky organic molecule from the resin matrix by shrinking the volume of the resin bead, and in the other instance, the chloride content of the brine displaces the organic matter components that are bonded to the resin by the ion exchange mechanism. Thus the concentration of brine, the quantity of brine used and the TOC present in the brine will determine the effectiveness of regeneration.

Figure 5 shows the capacity and breakthrough results for the recent pilot using a macroporous acrylic anion resin. One cubic foot of resin (28 liters) was used with a flowrate of 2 gallons per minute (16 BV/h). Organic matter in the supply water was quite high at an average of 11.3 ppm measured as total organic carbon (TOC). The test pilot was operated for approximately 40 days with roughly a 50% on-time factor. The cumulative percentage leakage is also shown. After approximately 4,000 bed volumes of capacity (30,000 gallons per cubic foot of resin), the percentage removal of TOC was still higher than 80%.

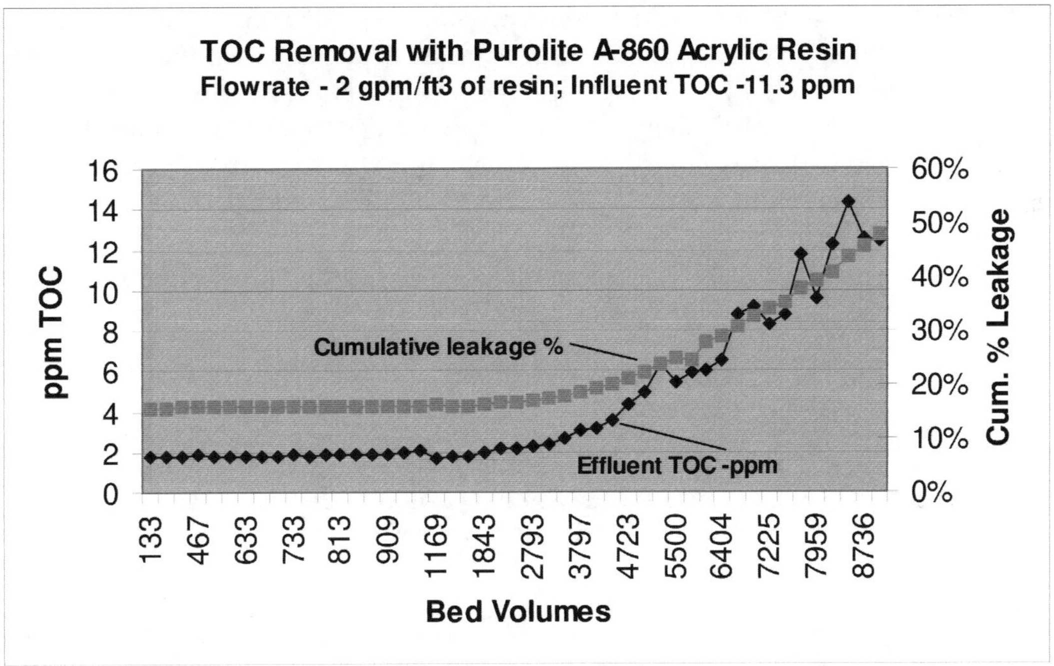

Figure 5. Typical Removal Efficiency of Acrylic SBA Resin

2.3 Brine Reuse

For the brine reuse evaluation, the pilot run was repeated using fresh resin for a service period of 5 weeks corresponding to the first breakpoint as discussed above, treating a total of 4076 bed volumes of water. The resin sample was then thoroughly mixed to achieve homogeneity and then separated into six equal fractions. Two regenerants were evaluated (a) 10% NaCl and (b) 10% NaCl + 2% NaOH. Three of the resin fractions were reserved for coflow regeneration with 1.3 bed volumes of 10% NaCl while the other three were reserved for coflow regeneration with 1.3 bed volumes 10% NaCl + 2% NaOH, both at a flow rate of 2 bed volumes per hour. For each set of resin fractions, the waste regenerant from the first fraction was subsequently used to regenerate the second resin fraction and the waste regenerant from the second used to subsequently regenerate the third fraction of resin. On each reuse of the regenerant, the first 25% of the regenerant was discarded and replaced by an equal volume of fresh regenerant in preparation for the next regeneration.

The initial and final TOC content of the resin fractions were measured and used as the measure of efficiency of regeneration of the resin. Initial TOC loaded on the resin was determined to be 31.3 mg as Carbon per dry gram of resin. Results are shown in Figure 6. Regeneration efficiency for both regenerants and for all regenerations ranged from approximately 87% for the first regeneration to approximately 82% for the third regeneration. Compared to regeneration efficiency for other common ion exchange processes which can typically range from 60% to 75%, all regeneration efficiency values obtained here are considered as excellent. The moderate reduction in efficiency from cycle to cycle indicates a good stabilization after three regenerations and it is deduced that the 25% regenerant "top-up" that was done in preparation for the next regeneration cycle was adequate to maintain the regeneration efficiency at these high levels. As a result, it is concluded that the brine reuse in conjunction with the "top-up" procedure used above can be easily extended to a minimum of five cycles before the regenerant must be considered for replacement. Also for this water quality, the lower cost option of regenerating with a simple 10% brine solution is adequate. Based on the above reuse strategy, the dosage of salt was reduced from 10 lbs/ft3 (160 g/l) for single use regeneration to a much lower value of 2.6 lbs/ft3 (42 g/l), amounting to approximately 75% savings in regenerant costs and environmental impact.

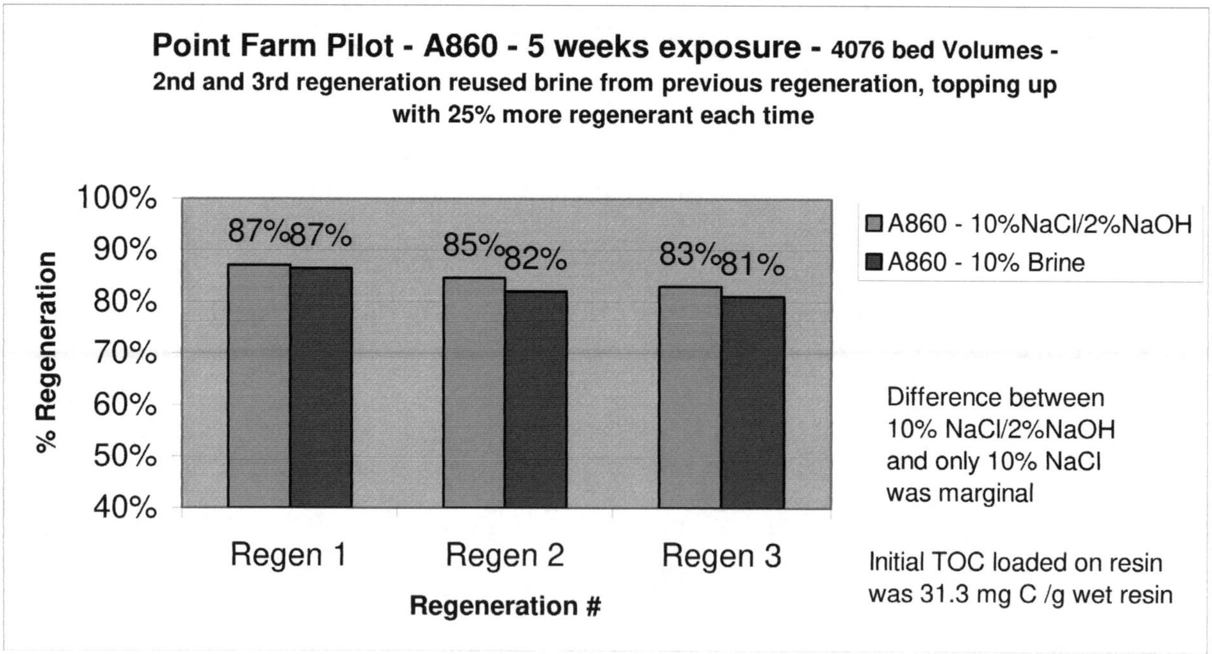

Figure 6. Brine Reuse using 10% NaCl or 10% NaCl + 2% NaOH

3.0 Cost Estimates

To complete the picture, an estimate of the operating cost for a single versus multiple brine use ion exchange system is needed. If a general resin price of $200 per cubic foot ($7/liter) is assumed along with a useful resin life of 5 years, the allocated cost of the resin to treat 1000 gallons (or $1m^3$) of water can be estimated. For example, assume a flowrate of 100 gpm (22.7m3/h) and a daily water production of 72,000 gallons (273 m^3). Resin volume needed at 2 gpm/ft^3 (16 BV/h) would be 50 cubic feet ($1.4m^3$), with allocated resin cost to treat the water estimated at approximately $0.04 per 1000 gallons ($0.01 per m^3). Assuming a salt cost of $0.05 per lb ($0.11/kg), a salt dosage of 8 lbs per cubic foot of resin (128g/l), and a working resin capacity of 8,000 gallons per cubic of resin (1069 BVs), the operating cost for salt would amount to $0.05 per 1000 gallons of water ($0.013 per m^3) for the single brine use system. For the multiple brine reuse system, the salt cost for the same condition would equate to $0.0125 per 1000 gallons of water ($0.0033/$m^3$). Table 1 compares the cost of salt and the amortized cost of the resin for the single use versus multiple use of the brine.

Table 1. Comparing Cost of Single Brine Use to Multiple Brine Reuse for TOC Removal

	Single Brine Use $ / 1000 Gals treated	Single Brine Use ($ / m3)	5 Reuses of Brine with 25% top-up /regeneration $ / 1000 Gals treated	5 Reuses of Brine with 25% top-up /regeneration ($ / m3)
Resin cost (5 yr life)	0.04	0.01	0.04	0.01
Salt Cost ($0.05 /lb $0.11/kg)	0.05	0.013	0.0125	0.0033
Total salt & resin cost (No bypass)	0.09	0.023	0.053	0.013

Table 2 attempts to show how ion exchange compares to other technologies, using cost data available from various sources. In the case of ion exchange, only the cost of salt and resin is shown.

Table 2. Comparing TOC Removal Cost by Various Technologies

	TOC Removal %	Capital Cost $	Operating Cost $/m3	Operating Cost $ / 1000 Gals
Acrylic SBA with brine reuse	> 50	N/A	0.013 (salt & resin only)	0.053 (salt & resin only)
MIEX [2]	> 50	$ 19 MM	0.026 – 0.04	0.10 – 0.15
GAC [2] Based on $1/lb GAC	25 - 50	$ 17 MM	0.095	0.36
MF + Coag.	10 - 25	N/A	0.09	0.35

4.0 Summary:

In summary, ion exchange technology provides a simple and cost effective method of controlling the formation of disinfection byproducts by efficiently removing organic matter before the point of disinfection of the water.

While single use of the brine regenerant is competitive for cases where brine discharges are still allowed, multiple recovery and reuse of the brine regenerant is efficient and can reduce operating cost for salt by over 75%.

If waste brine must be hauled away for disposal, multiple reuse of the brine before disposal can significantly reduce overall operating cost for TOC removal.

References:

1. "The removal of DBP precursors by GAC adsorption' by AWWA Research Foundation -1998

2. http://www.miexresin.com/ " MIEX Treatment versus GAC for TOC Removal" 2006

3. Clifford, D.A., Water Quality & Treatment Handbook, 4th Edition, Chapter 9, 1990.

Author Information:

Francis Boodoo, Technical Sales Manager for The Purolite Company, Bala Cynwyd, PA, has over 25 years experience in industrial, commercial and potable water treatment. Mr. Boodoo holds a Bachelor's degree in chemical engineering and a Master's degree in business administration and can be reached at fboodoo@puroliteusa.com.

DJ Shannahan, President of Sharp Water, has over 18 years experience in the water treatment and conditioning field as well as 11 years experience in chemical water treatment. Mr. Shannahan is a licensed water and wastewater plant operator in Maryland and Delaware and is a member of the Board of Directors of the Water Quality Association. He can be reached at djshannahan@sharpwater.com.

Edward Begg, Eastern Region Technical Sales Manager for the Purolite Company for, has over 30 years in industrial water treatment. Mr. Begg holds a Bachelor of Science degree in Chemistry from Long Island University and a Masters of Science in Environmental Engineering from Drexel University. He can be reached at tbegg@puroliteusa.com.

Joe D'Alessandro, North America Quality Manager for The Purolite Company, Bala Cynwyd, PA, has been with Purolite for over 17 years and has 32 years experience in polymer and analytical chemistry. He can be reached at jdalessandro@puroliteusa.com

307f Enhanced Chlorinated Organic Removal from Water Using Iron and Bimetallic Nanoparticle Composite Membranes

Stephen M. Ritchie, University of Alabama, Department of Chemical Engineering, Box 870203, Tuscaloosa, AL 35487-0203

Iron and bimetallic nanoparticles have shown immense promise for dechlorination of trichloroethylene (TCE) in water. However, application of these materials is not simple due to problems with material synthesis and oxidation during storage. In this work, iron and bimetallic nanoparticles are immobilized in a thin, polymeric membrane which combines absorption of the organic compound and dechlorination. This paper with discuss material synthesis and application of iron, iron-nickel, and iron-palladium membrane-supported nanoparticles for TCE dechlorination.

353a Investigation of Quality Concerns Due to Elastomeric Compunds Degradation in Chloramine-Disinfected Waters

Raja Mohan Nagisetty, Thomas D. Rockaway, and Gerold A. Willing. Civil and Environmental Engineering, University of Louisville, 2301 South 3rd Street, Civil and Environmental Engineering, Louisville, KY 40292

Due to the increasing health concerns and USEPA regulations, chlorine has increasingly been replaced by chloramine as the primary disinfectant in drinking water distribution systems. Even though, chloramines have been effective in reducing the disinfectant by-products, there have been reports of increasing elastomeric compound degradation. As the elastomer compound degradation progresses, it is noted that degradation by-products would leach into drinking waters and might lead to water quality concerns. To investigate this possibility, 30-day elastomer accelerated degradation tests were conducted for SBR, Natural Rubber and EPDM(S) elastomer samples in aqueous chloramine solutions and the solution samples were retrieved on 3, 6, 9, 15, 20 and 30 days for testing. The analytes were extracted using Solid Phase Micro Extraction (SPME) technique and the identification of organic compounds was carried out using Gas Chromatography/Mass Spectrometry. Varieties of carcinogenic organic compounds, leaching from the elastomer during the accelerated degradation tests, were identified and discussed in this study.

353b Adsorption of Chloroform & MTBE on Granular Zeolites; Batch and Column Adsorption Studies

Laila Abu-Lail, Civil & Environmental Engineering, Worcester Polytechnic Institute, 100 Institute Road, Worcester, MA 01609, John Bergendahl, Department of Civil & Environmental Engineering, Worcester Polytechnic Institute, 100 Institute Road, Worcester, MA 01609, Robert Thompson, Chemical Engineering, Worcester Polytechnic Institute, 100 Institute Road, Worcester, MA 01609, and Arjan Giaya, Triton Systems, Inc., 200 Turnpike Road, Chelmsford, MA 01824U.

Many powdered zeolites have been found to be effective at removing organics from water. But the passage of water through fixed beds of powdered zeolite produces high friction losses that are not present in flow through larger grain-sized, granular zeolites. In this work, batch and column adsorption of chloroform and methyl tertiary butyl ether (MTBE) onto granular zeolites was investigated. From results of the batch equilibrium experiments, it was found that the zeolite adsorbent with the greatest removal capacity for MTBE and chloroform from water was granular ZSM-5. Column adsorption experiments of MTBE & chloroform were performed using granular ZSM-5. Breakthrough curves obtained from these column experiments were used to predict the dynamics of fixed bed adsorbers for treating water. The breakthrough curves obtained from the column adsorption experiments were fitted to the bed depth service time (BDST) model.

353c Cost Estimation of Advanced Oxidation Processes (AOPs) Involving Ultrasound. A Comparative Study

Naresh N. Mahamuni[1], **Yusuf G. Adewuyi**[1], and Robert W. Peters[2]. (1) Chemical Engineering, North Carolina Agricultural & Technical State University, 1601 East Market Street, Greensboro, NC 27411, (2) University of Alabama at Birmingham, Department of Civil, Construction and Environmental Engineering, 1075 13th Street South, Birmingham, AL 35294-4440

Two things are needed for any technology to be suitable for use in the industry, viz.1.Technical feasibility and 2. Economical feasibility. The use of ultrasound for waste water treatment has been shown to be technically feasible by numerous reports in the literature over the years. But there are hardly any exhaustive reports which address the issue of economical feasibility of the use of ultrasound for waste water treatment on industrial scale.

Hence an attempt was made to estimate the cost for the waste water treatment using ultrasound. The costs have been calculated for 1000 lit/min capacity treatment plant. The costs were calculated based upon the rate constants for phenol degradation. The time required for ninety percent degradation of phenol was taken as the residence time. The amount of energy required to achieve the target conversion was calculated from the energy density (watt/ml) used in the treatability study. The cost of treatment was estimated by considering capital cost and operating cost involved for the waste water treatment. Quotations were invited from the vendors to estimate the capital cost of equipments involved, and operating costs were calculated based on annual energy usage. The cost is expressed in dollars per 1000 gallons of waste water treated. The cost has been estimated for phenol, TCE and azo-dye as model pollutants. These treatment costs are compared with other established AOP technologies. The cost of waste water treatment for phenol was in the range of $89 per 1000 gallons for UV/US/O3 to $15527 per 1000 gallons for US alone. These costs for TCE were in the range of $25 per 1000 gallons to $ 91 for US +UV treatment. The cost of waste water treatment for reactive dyes was in the range of $56 per 1000 gallon for US+UV+H2O2 to $2367 per 1000 gallon for US alone. Where possible, these estimated treatment costs are compared with AOP costs reported in the technical literature.

This study will help in quantifying the economics of waste water treatment using ultrasound on industrial scale. We strongly believe that this study will immensely help the researchers working in the area of applications of ultrasound for waste water treatment in terms of where the technology stands today as compared to other available commercial AOP technologies. This will also help them identify, for different ways to improve the efficiency of using ultrasound or search for other ways of generating cavitation which may be more efficient and help reduce the cost of treatment in the future.

353d Advanced Oxidation of Pharmaceuticals: Preparing for Indirect and Direct Water Reuse

Lee M. Blaney, Lynn E. Katz, and Desmond F. Lawler. Department of Civil, Architectural and Environmental Engineering, The University of Texas at Austin, 1 University Station C1786, Austin, TX 78712-0272

This research addresses two important issues associated with application of advanced oxidation processes (AOPs) for treatment of pharmaceutically active compounds (PhACs) in water reuse scenarios: composition of the background organic matter matrix and residual pharmaceutical activity (i. e., the post-treatment pharmacological activity associated with the target compound and/or its derivatives). Because AOPs are not selective, proper treatment design must consider the characteristic differences between human-impacted organic matter (HIOM), which derives from wastewater treatment plant effluent streams, and natural organic matter (NOM). Furthermore, because it is pharmacological activity, and not parent compound concentration, that threatens the contamination of drinking water resources, research efforts should focus on removing pharmacological activity.

In particular, this research focuses on advanced oxidation, via the peroxone (ozone/hydrogen peroxide) process, of two antibiotic drugs, namely, ciprofloxacin and gemifloxacin. Research tasks include the following: determine removal rates of target PhACs in the presence of varying background organic matter matrices derived from natural and reuse water sources; identify intermediate products that exhibit antibiotic activity; assess whether parent compound removal efficacy is directly proportional to eliminating residual pharmacological activity (RPA) or if intermediate products exert a significant fraction of RPA, which will be determined using a modified minimum inhibitory concentration (MIC) bioassay; develop a technique that can be employed to evaluate and optimize RPA removal from natural and impacted waters while minimizing pharmacologically active intermediates and advanced oxidation by-products.

The overarching goal of this research is development of a generalized strategy for assessing and predicting PhAC treatment efficacy, in terms of RPA removal, as a function of background organic matter matrix, treatment design, and operating parameters.

353e Removal of Volatile Organic Compounds from Water by Use of Microcapsules Containing PDMS

Ziying Xiang, Yangcheng Lu, Xingchu Gong, and Guangsheng Luo. Department of Chemical Engineering, The State Key Laboratory of Chemical Engineering,Tsinghua University, Beijing 100084, China

Volatile organic compounds(VOCs) in the wastewater have negative effects on environment and human health and development of efficient method to remove VOCs from water is an important task. In this study, poly(dimethylsiloxane) (PDMS) was found as a green extractant with large extraction capacity. However, because of its high viscosity, it is not suitable to be applied in liquid-liquid extraction directly. Therefore, PDMS was encapsulated in polyacrylonitrile microspheres as separation agent. The microspheres were prepared by solvent extraction method through a microfluidic device. The PDMS microcapsules had narrow size distribution and their mean diameter was about 500 μm. Microcapsules were employed to deal with wastewater containing low concentration of toluene, ethyl acetate and benzene respectively and the mass transfer performance of microcapsules was determined. Extraction equilibrium could be reached within 30min in batch experiment and microcapsules were reused for 4 times and no decline in the quantity of VOCs extracted was observed. The PDMS microcapsules can remove VOCs from water efficiently with no adverse coupling effects. A simple model was applied to batch experiments and the particle diffusion is probable the rate-controlling step of the process of extraction.

Water Compatible Resin for Separation and Recovery of Dissolved Precious Metals

Yu-Lung Lam, Die Yang, Chi-Yuet Chan, Kwong-Yu Chan, Patrick H. Toy

Chong Yuet Ming Chemistry Building, The University of Hong Kong, Pokfulam Road, HKSAR

Polystyrene-grafted-polyglycidol (PS-PG), a water-compatible resin, is applied for the extraction of precious metal in effluents of electroplating process. The extractions of gold (I) and silver (I) cyanide using this resin are investigated. More than 99% of gold and 99% of silver ions of the originally dissolved could be extracted from the aqueous phase to polymer phase. Recycling of the resins is possible and no deterioration in extraction performance was observed. Thiolation on the chemical structure shows improvement on the extraction capability and separation efficiency. The PS-PG group resin is a good option for metal extraction and separation because it is economical, environmental-friendly, non-toxic, non-flammable and inexpensive.

Preparation of materials for extraction

I. PS/PEG resins

PS/PEG resin was purchased directly from Rapp Polymere GmbH. The resin was used for extraction without any modification. The resin is called TentaGel S 30 900 with diameter of 90 micron.

II. Synthesis of polystyrene/polyglycidol (PS/PG) resins

This synthesis is a modification of R Haag *et al.*. In a 100 mL three-necked flask, hydroxymethylated magnetic beads (1 g, 7.1 mmol of OH group/g polymer) were suspended in dry diglyme (25 mL). After swelling for one hour, potassium t-butoxide (KO^tBu) (0.8 g, 7.1 mmol) in tetrahydrofuran (THF) was added at 40 °C over 12 h to deprotonate the hydroxyl groups on the PS beads. The mixture was heated to 120 °C and 50 mL dry dimethyformamide (DMF) was added. A solution of freshly distilled glycidol (24 mL, 26.4 g, 356 mmol) in dry DMF (40 mL) was added drop-wise to the reaction mixture. The mixture was stirred for 12 h at 120 °C and then quenched with 1 N hydrochloric acid (HCl). The mixture was filtered and washed with water and methanol (MeOH) and finaly dried in vacuum to give the product (21.6 g). The final white resins are composed of core polystyrene (4.64% by weight) covered with polyglycidol (95.36% by weight).

III. Synthesis of polystyrene/thiolated-polyglycidol (PS/PG-SH) resins

Mesyl chloride (1.0 g, 9 mmol) was added drop-wise to the mixture of PS-PG (0.5 g, 6.0 mmol of OH groups/g polymer) and pyridine (1.4 g, 18 mmol) in anhydrous methylene chloride (CH_2Cl_2). The reaction mixture was stirred for 12 hours at room temperature under an N_2 atmosphere. The resulting resin was filtered, washed with water and MeOH and then dried in vacuum to give the product 0.73 g of the product PS/PG-OMs.

Thiourea (1.6 g, 21 mmol) was added to the PS-PG-OMs (0.73 g) in ethanol (20 mL). The reaction mixture was refluxed overnight. After work-up, 0.48 g of functionalized polymer (PS-PG-SH) was obtained having a sulfur content of S 5.4 mmol/g.

IV. Synthesis of polystyrene/polyglycidol resins with a magnetite core (Mag/PS/PG)

Synthesis of polymer with magnetic core has been reported by many authors. Oleic acid, Fe(II) and Fe(III) were commonly used. In our experiment, magnetite nanoparticles were prepared by mixing 11.1 g $FeCl_3.6H_2O$ and 5.6g $FeSO_4.7H_20$ in 50 ml deionized (DI) water followed by addition of 20 ml of concentrated ammonia at 70°C. 1.5g oleic acid was added to the mixture and stirred for 30 minutes with sonification. The mixture was heated at 110oC for 30 minutes to remove water and excess ammonia. The synthesized magnetite nanoparticles were about 10 nm.

The oleic acid grafted magnetite nanoparticles were mixed together with 4-vinylbenzyl alcohol, divinyl benzene (DVB), and azobisisobutyronitrile (AIBN) in a three-necked flask with the mass concentrations of 0.48%, 4.26%, 0.54% and 0.09% respectively. The reaction mixture was stirred and degassed at room temperature under nitrogen (N_2) over 1 hour. DI water (94.54 wt %) with 0.09% dissolved polyvinyl alcohol (PVA) was also added to the mixture. Polymerization was performed for 16 hours at 70 °C. The PS-coated magnetite was washed with excess DI water and dried under vacuum.

In a 100 mL three-necked flask, hydroxymethylated PS-coated magnetic beads (1 g) were suspended in dry diglyme (25 mL), using methodologies well known in the art. After swelling for one hour, KO^tBu (0.8 g, 7.1 mmol) in THF was added at 40 °C over 12 hours to deprotonate the hydroxyl groups. The mixture was heated to 120 °C and 50 mL dry DMF was added. A solution of freshly distilled glycidol (24 mL, 26.4 g, 356 mmol) in dry DMF (40 mL) was added drop-wise to the reaction mixture. The mixture was stirred for 12 hours at 120 °C and then quenched with the addition of 1 molar HCl. The mixture was filtered and washed with water and methanol and finally dried in vacuum to give the product (3.5 g) with weight increase of 250%.

VI. Extraction Experiments

A stock solution of 15ppm Au was prepared from potassium gold (I) cyanide. Another stock solution of 15ppm Ag was prepared from potassium silver (I) cyanide. They were then used for metal extraction. For each metal extraction, 10 ml solution was mixed with 0.5g polymer. After stirring, the polymer was settled by sedimentation or magnetic separation and the aqueous phase was decanted for metal concentration determination.

One of the methods to determine the metal concentration is by Inductively Coupled Plasma – Atomic Emission Spectroscopy (ICP-AES). ICP-AES consists of a source of plasma. The plasma is generated in a quartz torch. Around the torch is a two or three turn induction coil, to which radio-frequency (RF) energy is applied. Argon is introduced into the torch and the RF-field is switched on. A magnetic field around the coil is developed and induces an electric field in the coil region. When seed electrons and ions are introduced with a Tesla discharge, plasma will be formed. The temperature of the plasma can be as high as 10000k which is sufficient to excite the electron to upper energy level as the sample flow through the plasma. As the lifetime of the excited atom is brief, its return to the ground state is accompanied by the emission of a photon of radiation. Since the emission line is characteristic to different metals, the intensity of the emission line is related to the concentration of the ion dissolved.

Results and Discussion

The obtained results indicated that gold can be extracted from the aqueous phase to the polymer phase. The determination of gold in aqueous phase after extraction showed the percentage of gold extracted to the polymer phase was higher than 99%. For silver extraction, the percentage of silver extracted to the polymer phase was 99%.

Recovery of gold and silver

After extraction, gold can be extracted reversely and recovered by means of electro-deposition as illustrated in Figure 1.

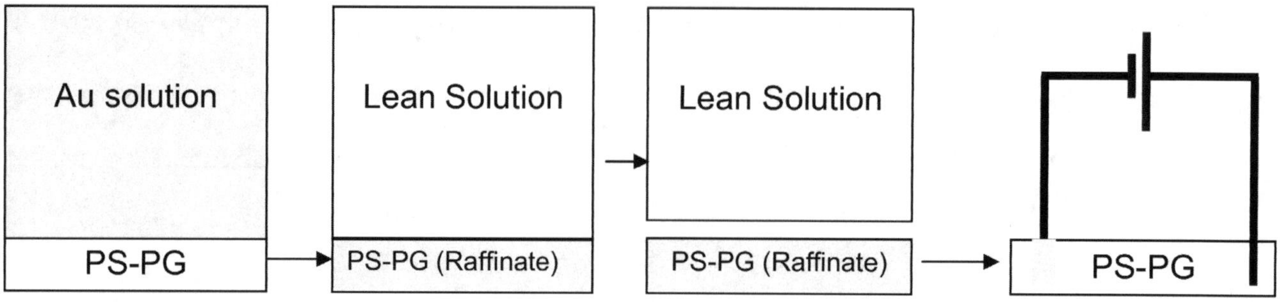

Figure 1. Schematic illustration of gold recovery

First, the gold ions in the water phase were extracted to the polymer phase so that resin saturated with gold was obtained. Secondly, the aqueous phase was discarded and only the resin phase (raffinate) was remained. Finally, a platinum wire as anode, the gold was plated on a clean nickel or copper wire. The electrical resistance in the polymer was very high and this caused high electrical voltage. The voltage could be reduced by adding small amount of salt such as sodium chloride or sodium sulfate.

The nickel wire was weighted before and after electro-deposition so that the mass of gold plated could be calculated. Similarly, silver recovery was also performed. From the experiment, though only about 40% to 50% of gold or silver were recovered from the resins, no deterioration of extraction capability on the resins was observed.

Enhancement for the extraction capability

In order to enhance the extraction performance of the polymer, a functional group was introduced into the PS/PG copolymer. The functional group incorporated is thiol group in which some of hydrogen atoms in the polymer are replaced by sulfur atoms. The lone pair electrons in the sulfur atoms have strong affinity towards metal ions as metal ions are generally electron deficiency. Therefore, metal-sulphide bond has been regarded as strong bonding and promote the extraction performance.

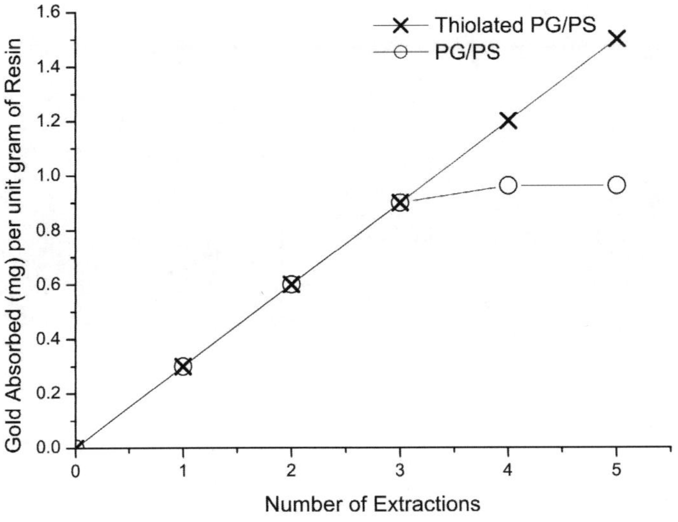

Figure 2. Comparsion between thiolated PS-PG and normal PS-PG on gold extraction capability

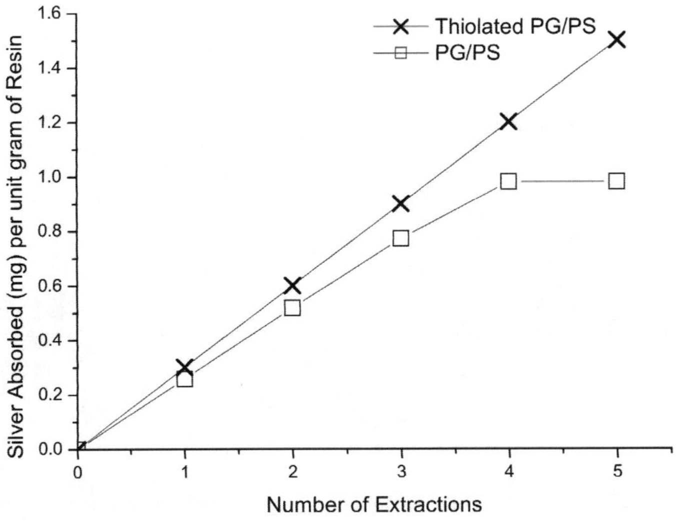

Figure 3. Comparsion between thiolated PS-PG and normal PS-PG on silver extraction capability

Gold could be reversely extracted from the thiolated polymer by means of sodium cyanide. A range of sodium cyanide with different concentrations was used to extract the gold from the thiolated resin saturated with gold reversely.

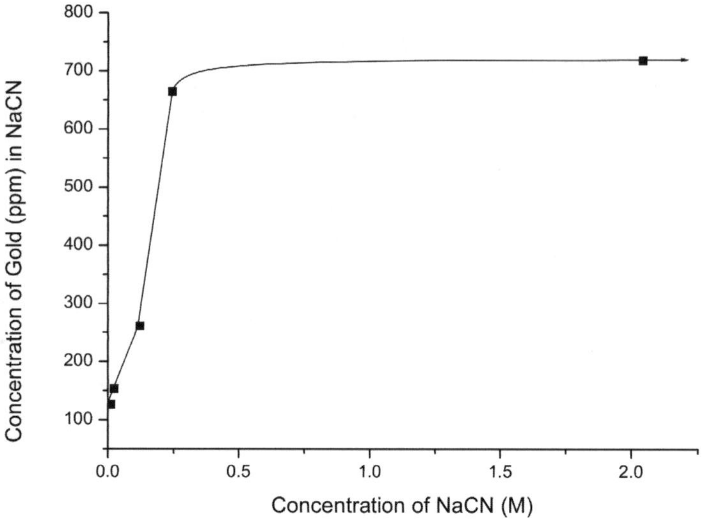

Figure 4. Effect of concentration of sodium cyanide on the gold reversely extracted from the thiolated PS-PG resins.

From the Figure 4, the optimum sodium cyanide concentration for gold reverse extraction is about 0.5M as higher concentration didn't provide further enhancement.

Conclusions

In conclusion, a polystyrene-grafted-polyglycidol (PS-PG) resin synthesized and was used for gold and silver extraction. The resin can be recycled without deterioration in extraction capability. In addition, significant increase in loading could be accomplished by introduction of thiol groups to the polymer. The metal re-extraction from the polymer phase is possible and can be made by electro-deposition or sodium cyanide leaching.

PS-PG can be a good candidate or replacement for polyethylene glycol in polyethylene glycol-based aqueous biphasic systems (PEG-ABS) as no addition of salt is required. Therefore, the system becomes less complicated and more convenient and easier for investigation and study.

Acknowledgement

This research is a joint project between the University of Hong Kong and Kenlap PGC Manufacturer Company Limited and has been supported by an award from the Innovation and Technology Fund. (UIM/166)

AUTHOR INDEX

Abu-Lail, Laila 88
Acheson, Carolyn 23
Adewuyi, Yusuf G. 67, 89
Athanassiou, Evagelos K. 45
Bachas, L. 3
Banavali, Rajiv 38
Begg, Edward T. 76
Bergendahl, John 64, 88
Bhattacharyya, D. 3
Blaney, Lee M. 90
Blute, Nicole 37
Boodoo, Francis 76
Brown, Derick G. 65
Chan, Kwong Yu 92
Chatterjee, Prasun K. 44
Dadwhal, Megha 25
D'Alessandro, Joe 76
Economy, James 66
Giaya, Arjan 88
Gong, Xingchu 91
Grass, Robert N. 45
Gu, Bachua 24
Gutierrez, Leonardo A. 27, 66
He, Feng 4
Höll, Wolfgang H. 16
Hug, Stephan J. 15
Hutchins, Steve 23
Kaegi, Ralf 15
Katsoyiannis, Ioannis A. 15
Katz, Lynn E. 90
Kawamura, Mitsuo 46
Kilduff, James 2
Koehler, Fabian M. 45
Lam, Yu Lung 92
Lawler, Desmond F. 90
Li, Xuan 27, 66
Lim, Hyung-Nam 2
Lin, Jin-Cheng 37
Lu, Yangcheng 91
Luo, Guangsheng 91
Mahamuni, Naresh N. 89
Martin, Todd 23
Méndez, José René Rangel 28
Meyer, David 3
Mills, Marc 23
Moller, Teresia 34
Mulchandani, Ashok 5
Nagisetty, Raja Mohan 87
Nerenberg, Robert 6
Nguyen, Thanh H. 27, 66
Padmanabhan, Anita 64
Parker, Garth R. 38
Peters, Robert W. 67, 89
Reible, Danny D. 1
Reyes, Refugio Bernardo García 28
Ritchie, Stephen M. 86
Roberts, Christopher B. 4
Roberts, Linda C. 15
Rockaway, Thomas D. 87
Rossier, Michael 45
Sahimi, Muhammad 25
Sarkar, Sudipta 37
Sayles, Gregory 23
Schenk, Kathleen 23
Sengupta, Arup K. 33, 37, 44
Shannahan, Dj 76
Sharma, M. P. 67
Sirkar, Kamalesh K. 22
Stark, Wendelin J. 45
Sylvester, Paul 34
Tee, Yit Hong 3
Thompson, Robert 88

AUTHOR INDEX

Toy, Patrick H 92
Trejo, Jose Antonio 38
Tsotsis, Theodore T. 25
Tsuji, Harue 46
Tsuji, Masamichi 46
Voegelin, Andreas 15
Wang, Jinwen 27
Willing, Gerold A. 87
Xiang, Ziying 91
Xu, Jian 3
Xu, Lucida 4
Young, Douglas 23
Zhao, Dongye 4

American Institute of Chemical Engineers
3 Park Avenue
New York, New York 10016-5991

CURRAN ASSOCIATES INC
proceedings
.com